TensorFlow移动端 机器学习实战

王众磊　陈海波 / 著

電子工業出版社
Publishing House of Electronics Industry
北京•BEIJING

内 容 简 介

随着人工智能技术的普及和边缘计算等新兴技术的兴起，很多人工智能的应用逐渐从云端向边缘设备和终端设备转移，基于移动端设备和嵌入式设备等小型设备的人工智能应用的开发越来越重要。

TensorFlow 作为开源机器学习框架，提供了对不同开发环境和设备的支持。本书详细讲解了如何使用 TensorFlow 进行端到端机器学习应用的开发，以及使用 TensorFlow Lite 在小型设备（包括 Android、iOS、树莓派（Raspberry Pi））上进行应用开发的要点和相应的实战案例。

本书也讲解了针对 Android 的硬件加速技术，以及业界流行的机器学习应用框架。本书代码基本对应 TensorFlow 2.0。

本书适合没有人工智能开发经验的初学者，以及有一定相关经验并且希望在人工智能应用上更加深入了解的开发者阅读。

未经许可，不得以任何方式复制或抄袭本书之部分或全部内容。
版权所有，侵权必究。

图书在版编目（CIP）数据

TensorFlow 移动端机器学习实战 / 王众磊, 陈海波著. -- 北京：电子工业出版社，2019.10

ISBN 978-7-121-37426-5

Ⅰ. ①T… Ⅱ. ①王… ②陈… Ⅲ. ①机器学习 Ⅳ. ①TP181

中国版本图书馆 CIP 数据核字(2019)第 200762 号

责任编辑：董 英
印　　刷：天津千鹤文化传播有限公司
装　　订：天津千鹤文化传播有限公司
出版发行：电子工业出版社
　　　　　北京市海淀区万寿路 173 信箱　邮编：100036
开　　本：787×980　1/16　印张：17　字数：340 千字
版　　次：2019 年 10 月第 1 版
印　　次：2019 年 11 月第 2 次印刷
定　　价：79.00 元

凡所购买电子工业出版社图书有缺损问题，请向购买书店调换。若书店售缺，请与本社发行部联系，联系及邮购电话：（010）88254888，88258888。

质量投诉请发邮件至 zlts@phei.com.cn，盗版侵权举报请发邮件至 dbqq@phei.com.cn。

本书咨询联系方式：010-51260888-819，faq@phei.com.cn。

前　言

2018年，我有很长一段时间在中国和美国两地跑，同时在国内工作和生活了比较长的一段时间，这是我近二十年来第一次和国内的开发者一起长时间工作。在享受各种美食之外，对国内的开发、产品和管理有了全新的了解和认识。

说起写书的起源，我本来的想法只是写一点可以作为国内工程师培训教材的东西。2018年初，TensorFlow作为一个技术热点，逐渐普及到机器学习应用开发的各个方面，但是对于TensorFlow在移动端的开发和应用还处于初始阶段。我当时也刚刚结束一个TensorFlow项目，想把这些经验和想法沉淀一下。于是我就把以前写的笔记和日志重新整理，添加一些内容并修改了文字，基本形成了一个原始版本。

后来，遇到博文视点的南海宝编辑，通过商谈，出版社欣然同意把这些资料整理出书。我的笔记和日志的内容很多和代码紧密相关，其中很多内容后来演变成了文档，我觉得这对初学者和有经验的开发者都是一个很好的参考，至少可以提供另外一个视角，让开发者多方面了解TensorFlow。所以，我就开始写作，前后花费了近两年的时间。

我是一边写作一边工作的，在这个过程中很快就遇到了两个很大的挑战。

第一是文字。我的笔记都是英文的，要把这些转换成中文，我借助了谷歌翻译，虽然翻译后的文字有很多需要修改，但至少省下了不少打字的时间。另外，就是专有术语的翻译，由于我对中文的专业术语不熟悉，所以即使简单的术语也要斟酌确定，这也花费了一些时间。如果读者在文字中发现一些奇怪的说法，还请见谅，我和编辑虽然尽了最大的努力，可能还是会有很多遗漏。

第二是重新认识和了解了国内开发的方方面面。我在美国和国内的开发者也有不少接触，我想在两边工作应该不会有什么差别，可实际工作起来还是有很多不同和挑战，感触颇深。首先是技术层面。开源的理念和软件在国内渗透到各个方面，几乎所有互联网公司都是从使用开源软件开始搭建自己的产品。由于谷歌在开源社区的贡献和影响力，国内普遍对谷歌的好感度很高，我也同享了这个荣耀。而且，很多公司和开发者也把对开源社区做出贡献看作责任和荣耀，这是一个很好的趋势，中国很快会发展出自己的开源生态和社区。

关于开发环境和工程师文化，我想提一下两边对新员工培训的区别。在国内对新员工的培训中，职业道德培训和公司文化的培训占了很大一部分。而在硅谷，至少像谷歌、脸书这些公司，培训中技术培训占了很大一部分，基本是一周的培训后，员工就要进行实际的工作，而国内很多公司的新员工第二周才开始技术工作。这里我能充分感受到中美公司之间的差别。

另外是开发管理方法，由于管理方法的不同，实际的工作中要做相应的改变。比如国内对开发和产品的进度的管理是非常严格的。但是，这种严格大都体现在层级的汇报关系上，而不是对技术细节的掌控和指导上。谷歌的工程师会经常以代码的提交作为一个工程开始和结束的标志，这在国内公司很少见到。

我希望把这些经验、想法和体会能或多或少体现在这本书里。比如，使用 Markdown 写文档，能使写文档变成一件不是很烦琐的事，可以让作者更专注于内容的写作，而不是花费太多时间在操作编辑器上。本书就是全部用 Markdown 写作完成，再转换成 Word 文档的。比如，使用 Bazel 编译，需要对代码的依赖有清晰的定义。可能很多工程师不会特别在意这点，但是通过它，工程师可以非常清楚地了解代码重用和引用的状况，避免随意的代码重用，并提高代码的质量。我希望通过这些在书中给读者传达一些不同的开发经验。

总之，我会把这本书作为 2018 年工作和生活的一个纪念。看到书中的各个章节，我就可以联想起写书时发生的许多事。但是，真的由于时间和我自己的能力非常有限，书中一定会有很多错误和瑕疵，还望读者能宽容和谅解。

最后，要感谢我的家人能支持和陪伴我度过 2018 年，我和我的母亲一起度过了 2018 年春节，是近 20 年来在国内度过的第一个春节。还要感谢我的妻子，她非常支持我，并帮助我写完这本书。还有我的两个女儿，总是能给我带来无尽的快乐，还要感谢深兰科技的创始人陈海波先生和首席战略官王博士，两位帮助我完成这本书，并提出了很多意见。

另外，感谢博文视点给我这个机会出版这本书，希望通过这本书能结识更多的开发者。还要感谢南海宝编辑在本书写作和出版过程中给予的指导和鼓励。

读者服务

扫码回复：37426

- 获取免费增值资源
- 获取精选书单推荐
- 加入读者交流群

目　　录

1

第 1 章 机器学习和 TensorFlow 简述

1.1 机器学习和 TensorFlow 的历史及发展现状

1.1.1 人工智能和机器学习

我们先来看一下人工智能（Artificial Intelligence）、机器学习（Machine Learning）、深度学习（Deep Learning）的定义。

人工智能（英文缩写为 AI）也称机器智能，指由人制造出来的机器所表现出来的智能。

下面是机器学习的英文定义：

Machine learning is a core, transformative way by which we're rethinking how we're doing everything.

其中文含义是：机器学习是一种核心的、变革性的方式，它正在改变我们思考的方式。人工智能（AI）是使事物变得聪明的科学，机器学习是一种开发人工智能的技术。

深度学习（Deep Learning）是机器学习的分支，是一种以人工神经网络为架构，对数据进行表征学习的算法。人工智能的分类如图 1-1 所示。

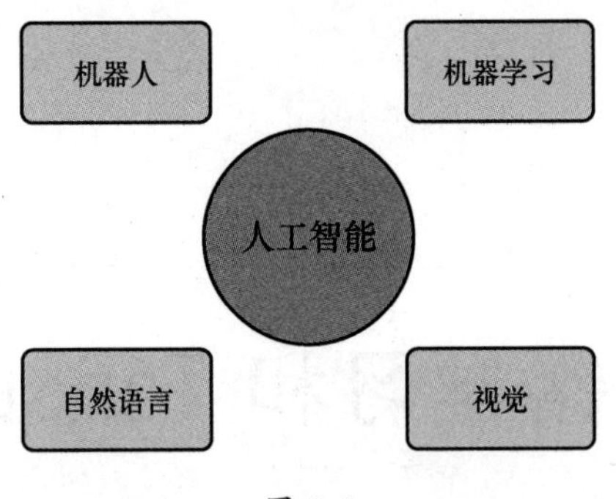

图 1-1

人工智能按产业分类大致可以分为下面几类：机器学习、自然语言处理、机器人技术和视觉等。在机器学习里深度学习是最近兴起也是比较热门的研究方面。自然语言处理和视觉的技术发展近几年来越来越成熟，有的技术已被大规模应用。

深度学习带来机器学习的革命。我们看到“深度学习”这个词在搜索中的热度近年来在快速攀升。arXiv 上的机器学习论文数量也在急剧增长。

深度学习（Deep Learning）是一种模仿人脑结构的机器学习。神经元是专注于某个特定方向的刺激（例如图像中对象的形状、颜色和透明度），通过将多个神经元分层组合在一起而完成模拟人脑的方法。分层可以模拟大脑运算，随着层数的增加，计算的功率和时间也会增加，进而提高计算的准确性。

下面来看一个图片分类的例子：给出一张图片，让机器识别这张图是一只猫还是一条狗。这个机器由多层的神经网络结构组成，该结构中有很多参数，经过大量训练之后，机器能识别出这张图是一只猫。如图 1-2（图片来源 https://becominghuman.ai/building-an-image-classifier-using-deep-learning-in-python-totally-from-a-beginners-perspective-be8dbaf22dd8）所示展示了这个深度学习的过程。

深度学习并不是全新的事物，但为什么在最近几年有了巨大突破？其中一个重要的原因是，人类发明了一个基于深度神经网络的解决方案。

在 20 世纪五六十年代，神经网络的研究就已经出现了，在马文 • 明斯基和西摩 • 帕尔特（1969）

发表了一项关于机器学习的研究以后，神经网络的研究就停滞不前了。

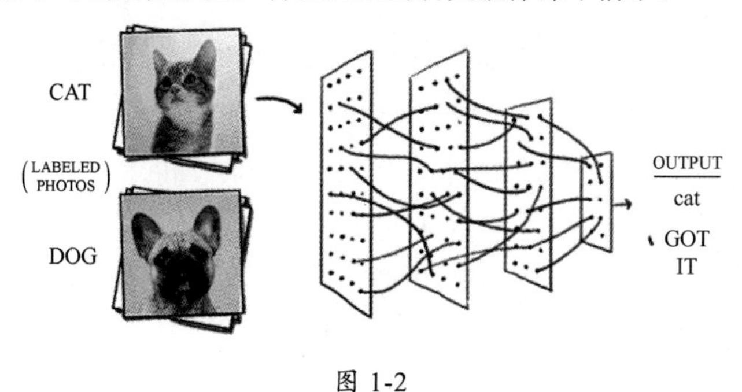

图 1-2

他们发现了神经网络的两个关键问题点。一个问题是，基本感知机无法处理异或回路；另一个问题是，计算机没有足够的能力来处理大型神经网络所需要的计算时间。在计算机具有更强的计算能力之前，神经网络的研究进展缓慢。但是随着计算能力的增加，深度学习解决问题的精度已经超过其他机器学习方法。

以图片识别为例，2011 年，机器识别的错误率是 26%，而人工识别的错误率只有 5%，所以这个时候的机器识别离实用有非常大的距离。到 2016 年，机器识别的错误率已经减少到 3% 左右，深度学习在该领域呈现出非常惊人的能力，这也是深度学习在图像识别领域吸引产业界大量关注的原因。

组成机器学习的三大要素是：数据、计算力和算法（Data、Computation and Algorithm）。

1.1.2　TensorFlow

接下来看一下最近非常流行的，也是本书主要讲解的机器学习框架 TensorFlow。

1. TensorFlow 的起源和发展历史

TensorFlow 是一个开源软件库，用于完成各种感知和语言理解任务的机器学习。TensorFlow 被 50 个团队用于研究和开发许多谷歌商业产品，如语音识别、Gmail、谷歌相册和搜索，其中许多产品曾使用过其前任软件 DistBelief。TensorFlow 最初由谷歌大脑团队开发，用于谷歌的研究和产品开发，于 2015 年 11 月 9 日在 Apache 2.0 开源许可下发布。

2010 年，谷歌大脑创建 DistBelief 作为第一代专有机器学习系统。谷歌的 50 个团队在谷歌和其他 Alphabet 公司的商业产品中部署了 DistBelief 的深度学习神经网络，包括谷歌搜索、谷歌语音搜索、广告、谷歌相册、谷歌地图、谷歌街景、谷歌翻译和 YouTube。

谷歌安排计算机科学家如 Geoffrey Hinton 和 Jeff Dean，简化和重构了 DistBelief 的代码库，使其变成一个更快、更健壮的应用级代码库，形成了 TensorFlow。

2009 年，Hinton 领导的研究小组通过在广义反向传播方面的科学突破，极大地提高了神经网络的准确性，使得神经网络的生成成为可能。值得注意的是，这个科学突破使得谷歌语音识别软件中的错误数减少了至少 25%。如图 1-3（横坐标为年份，纵坐标为谷歌搜索的人工智能和机器学习关键字数量）所示，2013 年以后，人工智能和机器学习的关键词搜索数量有了极大增长。

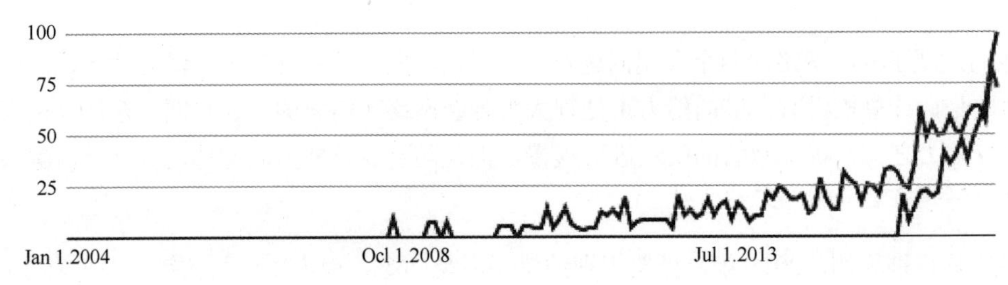

图 1-3

TensorFlow 是谷歌大脑的第二代机器学习系统。从 0.8.0 版本（发布于 2016 年 4 月）开始支持本地的分布式运行。从 0.9.0 版本（发布于 2016 年 6 月）开始支持 iOS。从 0.12.0 版本（发布于 2016 年 12 月）开始支持 Windows 系统。该移植代码主要是由微软贡献的。

TensorFlow 1.0.0 发布于 2017 年 2 月 11 日。虽然推理的实现运行在单台设备上，但 TensorFlow 也可以运行在多个 CPU 和 GPU（包括可选的 CUDA 扩展和图形处理器通用计算的 SYCL 扩展）上。TensorFlow 可用于 64 位的 Linux、macOS、Windows，以及移动计算平台（Android 和 iOS）。

TensorFlow 的计算使用有状态的数据流图来表示。TensorFlow 的名字来源于这类神经网络对多维数组执行的操作。这些多维数组被称为“张量”。2016 年 6 月，Jeff Dean 称：“在 GitHub 上有 1500 个库提到了 TensorFlow，其中只有 5 个来自谷歌。”

1.12.0 版本发布于 2018 年 10 月，TensorFlow 2.0 的预览版在 2019 年 3 月的 TensorFlow 开发者大会上发布。

2. TensorFlow 的主要特性

TensorFlow 开源以来已有 500 多个 Contributor 及 11000 多个 Commit。利用 TensorFlow 平

台在产品开发环境下进行深度学习的公司有 ARM、谷歌、UBER、DeepMind、京东等。谷歌把 TensorFlow 应用到很多内部项目，如谷歌语音识别、Gmail 邮箱、谷歌图片搜索等。

TensorFlow 有以下几个主要特性。

- 使用灵活：TensorFlow 是一个灵活的神经网络平台，采用图计算模型，支持 High-Level 的 API，支持 Python、C++、Go、Java 接口。
- 跨平台：TensorFlow 支持 CPU 和 GPU 的运算，支持台式机、服务器、移动平台的计算。并从 0.12 版本开始支持 Windows 平台。
- 产品化：TensorFlow 支持从研究团队发布模型到产品开发团队验证模型的全过程，构建起模型研究到产品开发实践的桥梁。
- 高性能：在 TensorFlow 中采用了多线程、队列技术及分布式训练模型，可在多 CPU、多 GPU 的环境下对模型进行分布式训练。

1.1.3　TensorFlow Mobile

其实 TensorFlow Mobile 并不是一个正式的名称，它只是 TensorFlow 对移动端设备和 IoT 设备的支持的总称，它可以直接移植到 Android、iOS 和树莓派等系统上。

1.1.4　TensorFlow Lite

2017 年 5 月谷歌宣布从 Android Oreo（API level 26）开始，提供一个专用于 Android 开发的软件栈 TensorFlow Lite。按照官方的定义，TensorFlow Lite 是为移动设备和嵌入式设备设计的机器学习软件框架，也是在移动和嵌入式设备上运行机器学习模型的官方解决方案。它支持在 Android、iOS 和其他操作系统上的低延迟和在较小二进制文件设备上的机器学习推理。

TensorFlow Lite 在 2017 年 10 月发布了第一个 Preview 版本，在官方网页 https://developers.googleblog.com/2017/11/announcing-tensorflow-lite.html 上有如下说明：

Lightweight Enables inference of on-device machine learning models with a small binary size and fast initialization/startup.

Cross-platform A runtime designed to run on many different platforms, starting with

Android and iOS.

Fast Optimized for mobile devices, including dramatically improved model loading times, and supporting hardware acceleration.

通过该段说明，我们可以将 TensorFlow Lite 的特性总结为如下三点：

（1）轻量级：使用小的二进制文件和快速初始化（启动），可以在设备端训练机器学习模型。

（2）跨平台：可以在 Android 和 iOS 的许多不同平台上运行。

（3）快速：针对移动设备进行优化，包括显著改进的模型加载时间和支持硬件加速。

如图 1-4（图片来源 https://techcrunch.com/2017/05/17/googles-tensorflow-lite-brings-machine-learning-to-android-devices）所示是 TensorFlow Lite 在谷歌 I/O 发布时的情景。

图 1-4

1.2 在移动设备上运行机器学习的应用

为什么要在移动端和 IoT 设备上进行机器学习？我们先来讨论下面几个问题：

- 为什么要在移动端上进行机器学习？
- 移动端上的机器学习要解决什么问题？
- 移动端机器学习面临的挑战是什么？

1.2.1　生态和现状

为什么要在移动设备和嵌入式设备上进行机器学习呢？在我国互联网的发展过程中，PC 互联网已经日趋饱和，移动互联网却呈现井喷式发展。中国互联网络信息中心发布的 2018 年互联网发展报告数据显示，截至 2018 年 6 月，中国手机网民超过 8.2 亿，占网民总数的 98.35%。随着移动终端价格的下降及 WiFi 的广泛使用，移动网民的数量呈现爆发趋势。

被称为“互联网女皇”的玛丽·米克尔在《2018 年互联网趋势报告》里说，中国的移动互联网行业在 2018 年迎来新的增长，中国网民人数已经超过 7.53 亿，占总人口的一半以上。移动数据流量消费同比上涨了 162%。

在智能手机市场，Android 和 iOS 是占比最大的两种操作系统，在全球已经激活的 31 亿智能手机中的占比超过 95%，而 Android 在两者之中又占据了绝对优势，截至 2017 年 11 月，Android 的市场份额为 75.9%，智能手机总计 23 亿部。中国和印度是全球最大的两个 Android 智能手机市场，占比接近一半。而根据 Newzoo 发布的《全球手机市场报告》，2018 年将再有 3 亿部新手机被激活，Android 手机的优势将进一步扩大。

Android 作为操作系统，它的生态包括各种类型的设备和仪器，从汽车、穿戴设备、VR/AR 设备到 IoT 设备，构成了一个庞大的系统，如图 1-5（图片来源 https://www.itproportal.com/features/iot-what-businesses-need-to-know/）所示。

在世界物联网博览会发布的《2017—2018 年中国物联网发展年度报告》中显示，2017 年全球物联网设备数量强劲增长，达到 84 亿台，首次超过人口数量。全球物联网市场有望在十年内实现大规模普及，到 2025 年市场规模或将增长至 3.9 万亿~11.1 万亿美元。

物联网发展呈现一些新的特点和趋势：一是全球物联网设备数量爆发式增长，物联网解决方案逐渐成熟；二是我国物联网市场规模突破万亿元，物联网云平台成为竞争核心领域；三是物联网细分领域热度出现分化，技术演进驱动应用产品向智能、便捷、低功耗方向发展。在 IoT 上运行的操作系统和应用，都要适用于物联网的特性。

图 1-5

1.2.2 从移动优先到人工智能优先

2017 年，谷歌决定将公司战略从移动优先转变为人工智能优先。虽然谷歌近十年来主要以移动优先，但很明显调整意味着公司看到了人工智能和机器学习技术的巨大潜力。2019 年，谷歌花了很大力气进行这种改变，并取得了显著的成果。

这里要说一下谷歌的 CEO Sundar Pichai，这位可算作硅谷最成功的印度裔的高管，一开始加入谷歌的 Chrome 团队，担任 Chrome 的产品经理。Chrome 是一款非常成功的产品，现在占据了浏览器市场超过 65%的份额，也是很多用户包括笔者自己的默认浏览器。

当然，ChromeOS 并不能算世界级的产品。虽然 Chrome 在谷歌内和其他世界级的产品相比并不是特别成功，但是它在教育领域及面向政府和企业的项目中或许会有令人瞩目的进展。有意思的是，很多一开始从事 ChromeOS 开发的工程师，后来开始谷歌 IoT 的研究，现在又在做新的操作系统的研究。

另外，谷歌在经历了几年移动优先之后，在 2017 年，Sundar Pichai 正式在谷歌 I/O 上提出了“从移动优先到人工智能优先”的战略。这是在继 ChormeOS 与 Android 整合后的一个非常重要的决定。我们可以看到，这几年人工智能已经成为一个非常热门的投资领域。

1.2.3　人工智能的发展

Jeff Dean 在人工智能发展刚刚取得突破性进展的时候就意识到，谷歌可能无法提供足够的计算力来支持人工智能的发展。后来，谷歌启动了 TPU 项目，通过硬件加速为人工智能提供强有力的计算基础。这体现在云计算和数据中心的 TPU POD 集群的部署中，也体现在对 Edge TPU 的研发和对移动端及 IoT 设备的进一步支持上。谷歌很重视基础技术和基础项目的研究，并愿意长时间地进行研发。后面，我们会介绍硬件加速的进展。如图 1-6（图片来源 https://www.datacenterknowledge.com/machine-learning/you-can-now-rent-entire-ai-supercomputer-google-cloud）所示是谷歌配备 TPU 的数据中心的照片。

图 1-6

1.2.4　在移动设备上进行机器学习的难点和挑战

在移动设备上进行机器学习是非常困难和具有挑战性的，主要表现在以下几个方面：

- 移动设备的 CPU 功率和电池电量非常有限。
- 设备和云之间的连接非常有限。
- 避免设备和云之间的大数据交换。

- 解决设备和云之间的遗留问题。
- 保护用户的隐私数据。

随着移动设备变得越来越强大，我们将看到更多运行在移动设备上的机器学习应用程序。

边缘计算是一个新技术。谷歌在 2016 年发布 TPU（张量处理单元），并在其数据中心大量使用 TPU。谷歌将 TPU 在数据中心的使用范围扩展到不同的域，将来我们或许可以运行完全支持 AI 的移动设备。

1.2.5 TPU

机器学习的发展一方面依赖软件和算法的提高，另一方面也离不开硬件的进步。2016 年 5 月，谷歌发布 TPU，一个专为机器学习和 TensorFlow 定制的 ASIC。TPU 是一个可编程的 AI 加速器，提供高吞吐量的低精度计算（如 8 位），用于使用或运行模型而不是训练模型。谷歌在数据中心运行 TPU 长达一年多，发现 TPU 对机器学习提供一个数量级更优的每瓦特性能。

2017 年 5 月，谷歌发布第二代 TPU，此款 TPU 在谷歌的 Compute Engine 中是可用的。第二代 TPU 提供最高 180 teraflops 的性能，在组装成 64 个 TPU 的集群时提供最高 11.5 petaflops 的性能。如图 1-7（图片来源 https://blog.hackster.io/announcing-the-new-aiy-edge-tpu-boards-98f510231591）和图 1-8（图片来源 http://hi.sevahi.com/google-unveils-tiny-new-ai-chips-for-on-device-machine-learning/）所示分别是谷歌对外公布的 TPU 和 Edge TPU 的图片。

图 1-7

图 1-8

1.3 机器学习框架

本节将介绍 TensorFlow Mobile、TensorFlow Lite 等工业界现有的机器学习框架，以及其他业界广泛使用的机器学习框架。另外，本节还将介绍对应移动设备和 IoT 设备的开发现状。

TensorFlow 具有不同的构建方式，并从一开始就支持不同的平台，例如 Android、iOS 和树莓派。值得一提的是，谷歌的物联网团队早期就是在树莓派上构建及运行 TensorFlow 的。

在本书中，我们主要讨论和描述如何在安卓设备上构建 TensorFlow。

TensorFlow 支持 TensorFlow Mobile 和 TensorFlow Lite 两个平台。本节将详细讨论两者的使用方法和区别，以及未来的发展趋势。

TensorFlow Mobile 有更好的支持全张量流的函数，我们可以在移动设备上运行张量流，但是 TensorFlow Mobile 并没有针对移动设备进行高度优化。

TensorFlow Lite 针对移动设备进行了高度优化，但作为一个新框架，它只具有有限的支持张量流功能。

1.3.1 CAFFE2

CAFFE（Convolutional Architecture for Fast Feature Embedding）是一个开源的深度学习框架，最初是在加州大学伯克利分校开发的。CAFFE 是用 C++编写的，支持 Python 接口。

贾清扬博士在加州大学伯克利分校攻读博士学位期间创建了 CAFFE 项目。现在该项目有很多贡献者，并在 GitHub 上进行托管。贾清扬博士现在是 Facebook 的人工智能工程总监，CAFFE 的服务框架也因此多多少少带上了 Facebook 的印记。

2017 年 4 月，Facebook 宣布了 CAFFE2，其中包括 Recurrent Neural Networks 等新功能。2018 年 3 月底，CAFFE2 被合并到 PyTorch。

CAFFE2 有不短的历史，对移动和嵌入式设备的支持也比较好。所以，很多在移动端进行机器学习开发的人会首选 CAFFE2。

1.3.2 Android NNAPI

Android NNAPI（Neural Networks API）是一个 Android C API，专门为在移动设备上对机器学习进行密集型运算而设计。NNAPI 旨在为构建和训练神经网络的更高级机器学习框架（例如 TensorFlow Lite、CAFFE2 或其他）提供一个基础的功能层。API 适用于运行 Android 8.1（API 级别为 27）或更高版本的所有设备。

1.3.3 CoreML

在苹果发布的 iOS11 中，有一个新的软件框架叫作 CoreML。使用 CoreML，可以将经过训练的机器学习模型集成到应用中。CoreML 是自然语言处理专业领域的框架和功能的基础。CoreML 支持用于图像分析的 Vision、用于自然语言处理的 Foundation，以及用于评估学习决策树的 GameplayKit。

CoreML 本身建立在低级原语之上，如 Accelerate、BNNS 及 Metal Performance Shaders。CoreML 针对器件内的性能进行了优化，可最大限度地减少内存占用和功耗。在设备上运行可确保用户数据的隐私性，并确保在网络连接不可用时，应用程序仍可正常运行并做出响应。想要了解更多的信息可以在苹果的开发者网页上进行查看。

以上苹果的这些描述是非常有意思的，这里提到了机器学习的特定技术，比如视觉、自然语音等，但是没有特别提到机器学习或人工智能，是一段比较严谨的描述。

随着人工智能和机器学习的发展，现在可选择的机器学习框架越来越多。比如由亚马逊、微软和英特尔共同开发现已成为 Apache 开源项目的 MXNet，由 NVIDIA 开发的可以实现硬件加速的 TensorRT，由阿里巴巴开源的深度学习框架 X-Deep Learning 和机器学习平台 PAI，以及由百度开源的深度学习框架 PaddlePaddle。这还不包括一些有着更长历史的机器学习框架，以及一些原生的计算框架。

作为普通开发者，在选择上会有一些难度。但是，由于机器学习和人工智能的复杂性，我们也看到了一些融合的趋势。比如，新的学习框架和老的学习框架的无缝融合和集成，学习框架对于 GPU、硬件和高计算的支持。开发者在选择框架的时候要考虑哪个框架适合自己，同时也要考虑这个框架的未来发展性。在 Wiki 里，有一个关于各种框架的比较分析，如图 1-9 所示。

Software	Creator	Software license[a]	Open source	Platform	Written in	Interface	OpenMP support	OpenCL support	CUDA support	Automatic differentiation[1]	Has pretrained models	Recurrent nets	Convolutional nets	RBM/DBNs	Parallel execution (multi node)	Actively Developed
Seq2SeqSharp	Zhongkai Fu	BSD	Yes	Windows	C#	C#	No	No	Yes	Yes	Yes	Yes	No	No	Yes	Yes
[illegible]	Kevin Lok	MIT license	Yes	Linux, macOS, Windows	Python	Python			Yes		Yes	Yes	Yes			
BigDL	Jason Dai	Apache 2.0	Yes	Apache Spark	Scala	Scala, Python			No		Yes	Yes	Yes			
Caffe	Berkeley Vision and Learning Center	BSD	Yes	Linux, macOS, Windows[2]	C++	Python, MATLAB, C++	Yes	Under development[3]	Yes	Yes	Yes[4]	Yes	Yes	No	?	
Deeplearning4j	Skymind engineering team; Deeplearning4j community; originally Adam Gibson	Apache 2.0	Yes	Linux, macOS, Windows, Android (Cross-platform)	C++, Java	Java, Scala, Clojure, Python (Keras), Kotlin	Yes	No[6]	Yes[6][7]	Computational Graph	Yes[8]	Yes	Yes	Yes	Yes[9]	
Chainer	Preferred Networks	MIT license	Yes	Linux, macOS, Windows		Python	No	No[10][11]	Yes	Yes	Yes	Yes	Yes			
Darknet	Joseph Redmon	Public Domain	Yes	Cross-Platform	C	C, Python	Yes	No[12]	Yes	Yes						
Dlib	Davis King	Boost Software License	Yes	Cross-Platform	C++	C++	Yes	No	Yes	Yes	Yes	No	Yes	Yes	Yes	
DataMelt (DMelt)	S.Chekanov	Freemium	Yes	Cross-Platform	Java	Java	No	No	No	No	No	Yes	Yes	Yes	Yes	
DyNet	Carnegie Mellon University	Apache 2.0	Yes	Linux, macOS, Windows		C++, Python		No[13]	Yes	Yes	Yes					
fastai library	fast.ai research lab Jeremy Howard Dr. Rachel Thomas	Apache License 2.0	Yes	Linux, macOS, Windows	Python	Python	?	No[14]	Yes[14]	?	Yes	Yes	Yes	?	?	Yes
Intel Data Analytics Acceleration Library	Intel	Apache License 2.0	Yes	Linux, macOS, Windows on Intel CPU[15]	C++, Python, Java	C++, Python, Java[15]	Yes	No	No	Yes	No		Yes		Yes	
Intel Math Kernel Library	Intel	Proprietary	No	Linux, macOS, Windows on Intel CPU[16]		C[17]	Yes[18]	No	No	Yes	No	Yes[19]	Yes[19]		No	
Keras	François Chollet	MIT license	Yes	Linux, macOS, Windows	Python	Python, R	Only if using Theano as backend	Can use Theano or Tensorflow as backends	Yes	Yes	Yes[20]	Yes	Yes	Yes	Yes[21]	
MATLAB + Neural Network Toolbox	MathWorks	Proprietary	No	Linux, macOS, Windows	C, C++, Java, MATLAB	MATLAB	No	No	Train with Parallel Computing Toolbox and generate CUDA code with GPU Coder[22]	No	Yes[23][24]	Yes[23]	Yes[23]	No	With Parallel Computing Toolbox[25]	Yes
Microsoft Cognitive Toolkit	Microsoft Research	MIT license[26]	Yes	Windows, Linux[27] (macOS via Docker on roadmap)	C++	Python (Keras), C++, Command line,[28] BrainScript[29] (.NET on roadmap[30])	Yes[31]	No	Yes	Yes	Yes[32]	Yes[33]	Yes[33]	No[34]	Yes[35]	Yes
Apache MXNet	Apache Software Foundation	Apache 2.0	Yes	Linux, macOS, Windows,[36][37] AWS, Android,[38] iOS, JavaScript[39]	Small C++ core library	C++, Python, Julia, Matlab, JavaScript, Go, R, Scala, Perl	Yes	On roadmap[40]	Yes	Yes[41]	Yes[42]	Yes	Yes	Yes	Yes[43]	
Neural Designer	Artelnics	Proprietary	No	Linux, macOS, Windows	C++	Graphical user interface	Yes	No	Yes	?	?	No	No	No	?	
OpenNN	Artelnics	GNU LGPL	Yes	Cross-platform	C++	C++	Yes	No	Yes	?	?	No	No	No	?	
PaddlePaddle	Baidu	Apache License	Yes	Linux, macOS, Windows	C++, Python	Python	No	Via the mobile based version[44]	Yes	Yes	Yes	Yes	Yes	?	Yes	
PlaidML	Vertex.AI	AGPL3	Yes	Linux, macOS, Windows	C++, Python	Keras, Python, C++, C	No	Yes	Yes	Yes		Yes	Yes	?	Yes	
PyTorch	Adam Paszke, Sam Gross, Soumith Chintala, Gregory Chanan	BSD	Yes	Linux, macOS, Windows	Python, C, CUDA	Python	Yes	Via separately maintained package[45][46][47]	Yes	Yes	Yes	Yes	Yes		Yes	
Apache SINGA	Apache Incubator	Apache 2.0	Yes	Linux, macOS, Windows	C++	Python, C++, Java	No	Supported in V1.0	Yes	?	Yes	Yes	Yes	Yes	Yes	
TensorFlow	Google Brain team	Apache 2.0	Yes	Linux, macOS, Windows,[48] Android	C++, Python, CUDA	Python (Keras), C/C++, Java, Go, JavaScript, R,[49] Julia, Swift	No	On roadmap[50] but already with SYCL[51] support	Yes	Yes[52]	Yes[53]	Yes	Yes	Yes	Yes	Yes
TensorLayer	Hao Dong	Apache 2.0	Yes	Linux, macOS, Windows,[48] Android	C++, Python	Python	No	On roadmap[50] but already with SYCL[51] support	Yes	Yes[52]	Yes[53]	Yes	Yes	Yes	Yes	
Theano	Université de Montréal	BSD	Yes	Cross-platform	Python	Python (Keras)	Yes	Under development[54]	Yes	Yes[55][56]	Through Lasagne's model zoo[57]	Yes	Yes	Yes	Yes[58]	No
Torch	Ronan Collobert, Koray Kavukcuoglu, Clement Farabet	BSD	Yes	Linux, macOS, Windows,[59] Android,[60] iOS	C, Lua	Lua, LuaJIT,[61] C, utility library for C++/OpenCL[62]	Yes	Third party implementations[63][64]	Yes[65][66]	Through Twitter's Autograd[67]	Yes[68]	Yes	Yes	Yes	Yes[69]	
Wolfram Mathematica	Wolfram Research	Proprietary	No	Windows, macOS, Linux, Cloud computing	C++, Wolfram Language, CUDA	Wolfram Language	Yes	No	Yes	Yes	Yes[70]	Yes	Yes	Yes	Under Development	
VerAI	VerAI	Proprietary	No	Linux, Web-based	C++, Python, Go, Angular	Graphical user interface, cli	No	No	Yes	Yes	Yes	Yes	Yes	Yes	Yes	

图 1-9　机器学习框架的比较

注：该图比较大，读者可到 https://en.wikipedia.org/wiki/Comparison_of_deep_learning_software 上查阅。

1.3.4　树莓派（Raspberry Pi）

依据树莓派网站 https://www.raspberrypi.org/products/raspberry-pi-3-model-b-plus 的介绍，它们的最新产品是 Raspberry Pi 3 Model B+。它配置有一个 4 核 64bit 的 Cortex-A53 处理器，主频 1.4GHz。同时也支持无线网、蓝牙等。秉承一贯的风格，树莓派的开发板小巧同时又能提供相当的计算能力。树莓派被用作很多 IoT 或嵌入式开发的参照板。机器学习需要在这种计算能力、储存能力等都有限的平台上表现出优越的性能，才能在更广阔的范围内得到推广和应用。

2

第 2 章 构建开发环境

这一章，我们将学习怎样构建 TensorFlow 的开发环境。在介绍构建 TensorFlow 特有的开发环境之前，我们还要介绍一些同类软件开发环境的构建方法。

2.1 开发主机和设备的选择

许多开发人员使用 Windows 作为开发平台，但 Android 不支持 Windows 作为开发平台。多数谷歌工程师使用定制的 Linux 作为主要平台，因为使用相同的平台可以避免任何可能由平台引起的角落案例问题。建议开发者安装 Ubuntu 16 或使用 macOS。熟悉 Linux 对于 Android 开发非常有帮助，因为 Linux 上的许多概念和命令工具都可以在 Android 上使用。

Android 开发人员可以在 AOSP（Android Open-Source Project，安卓开放源代码项目）

中开发应用程序，但这种开发方式非常复杂。因此，大多数开发人员选择 Android Studio 作为开发平台。

在此，笔者鼓励开发人员使用基于 AOSP 开发标准的独立应用程序。很多 OEM 和 ODM 都是从 AOSP 开始开发定制的 Android 的，但是，随着开发规模和代码规模的不断增长，开发人员应该考虑将平台开发和应用程序开发分开。通过这种方式，应用程序开发将只依赖于平台 SDK 而非源代码，并且应用程序可以做更快的开发迭代、测试和部署。

2.2　在网络代理环境下开发

在网络代理或防火墙后面进行开发，都是非常困难的。因为在国内，很多公司仍然使用 Windows 作为主要的开发平台，而且很多公司的 IT 部门对开源社区没有很好的支持。我们要逐一解决这些问题。对于代理人背后的开发者，开发人员必须正确设置 GitHub.com 等网站。

在网络代理环境下开发，会有很多事情需要处理。首先，需要正确设置代理、用户和密码，这种设置可能会分散到多个文件中，因此请尝试记下更改的内容及执行此操作的最佳方法。

接下来解决认证问题，即许多网站的识别用户功能。简单的方法可能是配置设置跳过或忽略身份验证，但大多数工具都有自己的设置，所以必须将它们分开配置。有些工具在内部会调用其他工具，在配置过程中需要确定哪个工具会出现问题。在开发过程中最坏的情况是：需要编写代码，重新构建工具或在工具中添加代理支持。为通过代理从 Maven 下载，笔者通过重建 Git 来支持 OpenSSL 和 Hack Bazel。

在 Ubuntu 16.04 上，笔者对以下工具参数进行了设置：

- .bashrc
- apt
- CNTLM
- git
- curl
- wget

- Bazel

具体设置参数和内容如表 2-1 所示。

表 2-1　具体设置参数和内容

设置参数	内　容	设置参数	内　容
.bashrc	proxy	curl	authentication
apt	proxy	wget	authentication
CNTLM	authentication	Bazel	maven download
git	authentication		

2.3 集成开发环境 IDE

2.3.1 Android Studio

Android Studio 是谷歌支持的官方 IDE。

下载并安装 Android Studio。Android Studio 支持 Gradle，但在 2018 年 6 月以前，3.1.2 版本不支持 Bazel 0.13。希望谷歌 Android Studio 团队和 Blaze 团队能够更好地同步并提供无缝支持，否则 Android 社区很难采用 Bazel。

2.3.2 Visual Studio Code

Visual Studio Code 是由微软发布的开发集成环境。其官方网站上写道：

Visual Studio Code 是一个轻量级但功能强大的源代码编辑器，可在桌面上运行，可用于 Windows、macOS 和 Linux。它支持 JavaScript、TypeScript 和 Node.js 等脚本，并且具有丰富的语言（如 C ++、C# 、Java、Python、PHP、Go）和运行时（如.NET 和 Unity）扩展生态系统。

Visual Studio Code 的安装步骤非常简单。安装完毕后，根据需要继续安装 Bazel 扩展和其他语言的扩展。如图 2-1 所示为 Bazel 的扩展页面。

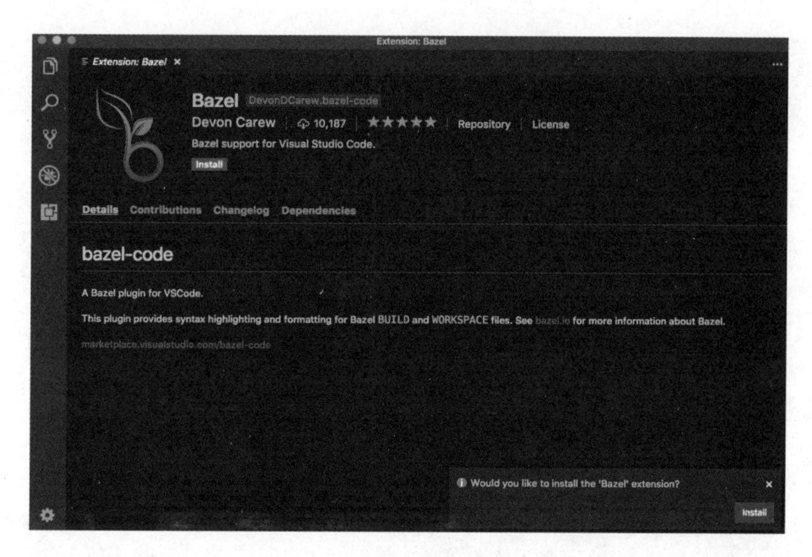

图 2-1　Bazel 的扩展页面

开发人员也可以安装一些 Android 插件，图 2-2 展示了安装 Bazel 插件的相关信息。

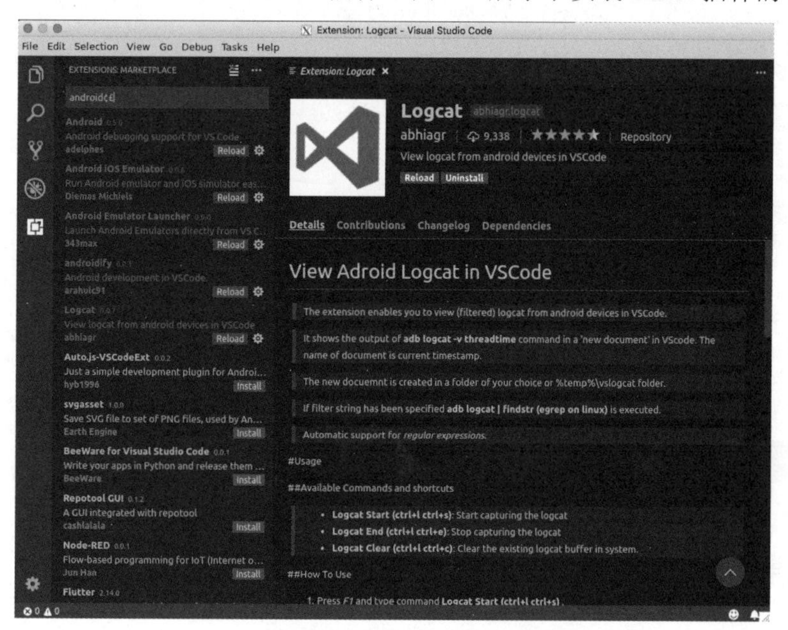

图 2-2　安装 Bazel 插件

安装完相关插件后，Visual Studio Code 应该正常工作，包括 C++/Java/Python 等语言支持功能和自动完成功能，图 2-3 为 Visual Studio Code 的编辑界面。

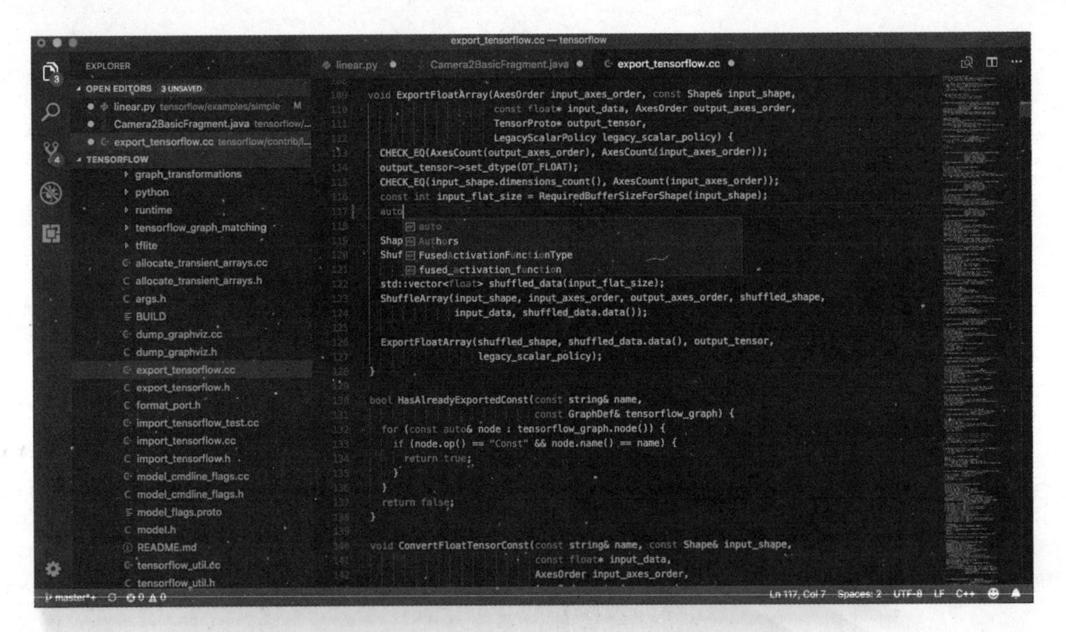

图 2-3 Visual Studio Code 的编辑界面

不得不说，微软在支持开源社区方面做得确实很好，这引起了很多开发者的关注。至少，笔者在写这本书的时候，使用的 IDE 就是 Visual Studio Code。

2.3.3 其他 IDE

除 Visual Studio Code 外，还有 IntelliJ、Eclipse 等多种工具供开发者选择使用， 开发者可根据个人习惯选择不同的工具。在过去很长一段时间内，Eclipse、Emacs 和 Vi 占据了 IDE 的主流市场，随后过渡到 IntelliJ，目前以 Android Studio 为主。还有另一个强大的基于云的 IDE 在谷歌中越来越受欢迎。

2.4 构建工具 Bazel

大多数开发人员使用 Android Studio + Gradle + Maven 作为他们的日常工具，这种配置已经足够了。不过，笔者强烈建议使用 Bazel，它是谷歌内部工具的开源版本。Bazel 具有很多重要的功能，最新版本是 Bazel 0.22，与 1.0 正式发布版本非常接近。很多离开谷歌的工程师，仍会继续使用 Bazel。由 Facebook 开发的 Buck 与 Bazel 的功能非常相似。许多初创公司使用的也是 Bazel 工具。

Bazel 的构建方法很简单，在 Linux 上执行下面的命令就可以安装 Bazel：

```
sudo apt-get install bazel
sudo apt-get upgrade bazel
```

Android Studio 也支持 Bazel。在 Android Studio 上安装 Bazel 插件即可使用，安装 Bazel 插件的步骤如图 2-4 所示。

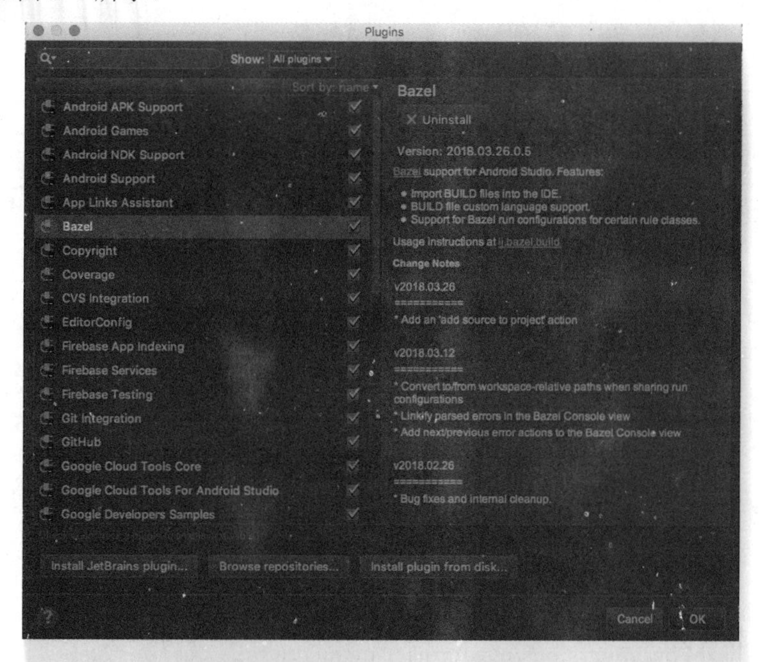

图 2-4　在 Android Studio 上安装 Bazel 插件

TensorFlow 中支持 Gradle、Make 和 iOS 的构建，谷歌内部不支持 Gradle，只是在开源的时候会对其进行测试。

2.4.1　Bazel 生成调试

生成调试命令为：

```
bazel build --compilation_mode=dbg //tensorflow/lite/toco
```

2.4.2 Bazel Query 命令

如果读者想找出 tensorflow/python/saved_model 和 tensorflow/core/kernels:slice_op 算子之间的关系，可以执行 bazel query "somepath(tensorflow/python/saved_model, tensorflow/core/kernels:slice_op)"，执行结果如下：

```
//tensorflow/python/saved_model:saved_model
//tensorflow/python/saved_model:builder
//tensorflow/python:lib
//tensorflow/python:pywrap_tensorflow
//tensorflow/python:pywrap_tensorflow_internal
//tensorflow/python:_pywrap_tensorflow_internal.so
//tensorflow/python:tf_session_helper
//tensorflow/core:all_kernels
//tensorflow/core/kernels:array
//tensorflow/core/kernels:slice_op
```

如果读者想找出//tensorflow/contrib/lite/java/demo/app/src/main:TfLiteCameraDemo 和//tensorflow/contrib/lite/kernels:builtin_ops 算子之间的依赖关系，可以执行 bazel query "somepath(//tensorflow/contrib/lite/java/demo/app/src/main:TfLiteCameraDemo,//tensorflow/contrib/lite/kernels:builtin_ops)"，执行结果如下：

```
//tensorflow/lite/java/demo/app/src/main:TfLiteCameraDemo
//tensorflow/lite/java:tensorflowlite
//tensorflow/lite/java:tflite_runtime
//tensorflow/lite/java:libtensorflowlite_jni.so
//tensorflow/lite/java/src/main/native:native
//tensorflow/lite/kernels:builtin_ops
```

除 Android 特有的设定外，我们还要做 TensorFlow 的标准设定，具体的步骤请参考网页 https://docs.bazel.build/versions/master/user-manual.html。

2.5 装载 TensorFlow

TensorFlow 的安装比较简单，官网上提供了详细的说明，具体内容请参阅 TensorFlow 的网上链接。

需要提醒的是，建议使用 Virtualenv 来安装 TensorFlow，安装完 TensorFlow 和 GPU（图形加速器）支持后，需要验证。

由于某些用户在安装 TensorFlow GPU 支持时会遇到问题，因此接下来将介绍如何安装

GPU 支持。

首先，开发者要在 Developer Nvidia Website 上注册。

然后，按照此链接安装 GPU。

接着，还需要安装 CUDA Toolkit 9.0，tensorflow.org 中的链接始终指向最新的 CUDA 版本，现在是 9.2 版本。但是不要使用 9.2 版本，除非 TensorFlow 支持它。请使用上面链接的 CUDA 9.0 版本。

同样，请下载并安装 cuDNN v7.1.4 for CUDA 9.0，tensorflow.org 中的链接指向的最新版 cuDNN 是 CUDA 9.2 的 v7.1.4 版本。安装并运行如下命令：

```
$ nvcc -V
nvcc: NVIDIA (R) Cuda compiler driver
Copyright (c) 2005-2017 NVIDIA Corporation
Built on Fri_Sep__1_21:08:03_CDT_2017
Cuda compilation tools, release 9.0, V9.0.176
```

接着，运行命令“$ nvidia-smi”，得到如下结果：

```
Fri Jun 15 22:21:08 2018
+-----------------------------------------------------------------------------+
| NVIDIA-SMI 384.130                Driver Version: 384.130                   |
|-------------------------------+----------------------+----------------------+
| GPU  Name        Persistence-M| Bus-Id        Disp.A | Volatile Uncorr. ECC |
| Fan  Temp  Perf  Pwr:Usage/Cap|         Memory-Usage | GPU-Util  Compute M. |
|===============================+======================+======================|
|   0  Quadro K600         Off  | 00000000:05:00.0 Off |                  N/A |
| 25%   48C    P0    N/A /  N/A |      0MiB /   979MiB |      0%      Default |
+-------------------------------+----------------------+----------------------+

+-----------------------------------------------------------------------------+
| Processes:                                                       GPU Memory |
|  GPU       PID   Type   Process name                             Usage      |
|=============================================================================|
|  No running processes found                                                 |
+-----------------------------------------------------------------------------+
```

此外，还要在 CUDA 示例代码中运行 deviceQuery，以确保 GPU 正常工作。

```
Device 0: "Quadro 600"
CUDA Driver Version / Runtime Version          9.0 / 9.0
```

```
  CUDA Capability Major/Minor version number:    2.1
  Total amount of global memory:                 962 MBytes (1009254400 bytes)
```

执行结果如下：

```
Quadro M6000
```

如果你看到类似的结果，说明你的显卡可以支持 TensorFlow。

然后，我们执行下面的命令：

```
$ ./bin/x86_64/linux/release/deviceQuery
```

执行结果会显示显卡的版本号和各种性能数据：

```
./bin/x86_64/linux/release/deviceQuery Starting...

 CUDA Device Query (Runtime API) version (CUDART static linking)

Detected 1 CUDA Capable device(s)

Device 0: "Quadro M6000 24GB"
  CUDA Driver Version / Runtime Version          9.0 / 9.0
  CUDA Capability Major/Minor version number:    5.2
  Total amount of global memory:                 24467 MBytes (25655836672 bytes)
  (24) Multiprocessors, (128) CUDA Cores/MP:     3072 CUDA Cores
  GPU Max Clock rate:                            1114 MHz (1.11 GHz)
  Memory Clock rate:                             3305 Mhz
  Memory Bus Width:                              384-bit
  L2 Cache Size:                                 3145728 bytes
  Maximum Texture Dimension Size (x,y,z)         1D=(65536), 2D=(65536, 65536),
3D=(4096, 4096, 4096)
  Maximum Layered 1D Texture Size, (num) layers  1D=(16384), 2048 layers
  Maximum Layered 2D Texture Size, (num) layers  2D=(16384, 16384), 2048
layers
  Total amount of constant memory:               65536 bytes
  Total amount of shared memory per block:       49152 bytes
  Total number of registers available per block: 65536
  Warp size:                                     32
  Maximum number of threads per multiprocessor:  2048
  Maximum number of threads per block:           1024
  Max dimension size of a thread block (x,y,z): (1024, 1024, 64)
  Max dimension size of a grid size    (x,y,z): (2147483647, 65535, 65535)
  Maximum memory pitch:                          2147483647 bytes
```

```
    Texture alignment:                             512 bytes
    Concurrent copy and kernel execution:          Yes with 2 copy engine(s)
    Run time limit on kernels:                     Yes
    Integrated GPU sharing Host Memory:            No
    Support host page-locked memory mapping:       Yes
    Alignment requirement for Surfaces:            Yes
    Device has ECC support:                        Disabled
    Device supports Unified Addressing (UVA):      Yes
    Supports Cooperative Kernel Launch:            No
    Supports MultiDevice Co-op Kernel Launch:      No
    Device PCI Domain ID / Bus ID / location ID:   0 / 4 / 0
    Compute Mode:
       < Default (multiple host threads can use ::cudaSetDevice() with device
simultaneously) >

  deviceQuery, CUDA Driver = CUDART, CUDA Driver Version = 9.0, CUDA Runtime
Version = 9.0, NumDevs = 1
  Result = PASS
  $ nvidia-smi
  +-----------------------------------------------------------------------------+
  | NVIDIA-SMI 384.130                Driver Version: 384.130                   |
  |-------------------------------+----------------------+----------------------+
  | GPU  Name        Persistence-M| Bus-Id        Disp.A | Volatile Uncorr. ECC |
  | Fan  Temp  Perf  Pwr:Usage/Cap|         Memory-Usage | GPU-Util  Compute M. |
  |===============================+======================+======================|
  |   0  Quadro M6000 24GB   Off  | 00000000:04:00.0  On |                  Off |
  | 25%   41C    P8    20W / 250W |    488MiB / 24467MiB |      0%      Default |
  +-------------------------------+----------------------+----------------------+


  +-----------------------------------------------------------------------------+
  | Processes:                                                       GPU Memory |
  |  GPU       PID   Type   Process name                             Usage      |
  |=============================================================================|
  |    0      2183      G   /usr/lib/xorg/Xorg                           319MiB |
  |    0      3796      G   compiz                                        92MiB |
  |    0      6095      G   ...-token=32ADD0D4261B4355966B2810A61BBF37    72MiB |
  +-----------------------------------------------------------------------------+
```

最后，还要安装 TensorFlow GPU：

```
(tensorflow)$ pip install --upgrade tensorflow      # for Python 2.7
(tensorflow)$ pip3 install --upgrade tensorflow     # for Python 3.n
(tensorflow)$ pip install --upgrade tensorflow-gpu  # for Python 2.7 and GPU
(tensorflow)$ pip3 install --upgrade tensorflow-gpu # for Python 3.n and GPU
```

安装成功之后，可以用下面的命令确认：

```
(tensorflow) $ python
Python 2.7.12 (default, Dec  4 2017, 14:50:18)
[GCC 5.4.0 20160609] on linux2
Type "help", "copyright", "credits" or "license" for more information.
>>> import tensorflow as tf
>>> hello = tf.constant("hello")
>>> sess = tf.Session()
**2018-06-20  06:54:34.284161:  I  tensorflow/core/platform/cpu_feature_
guard.cc:140] Your CPU supports instructions that this TensorFlow binary was not
compiled to use: AVX2 FMA**
**2018-06-20  06:54:34.460555:  I  tensorflow/core/common_runtime/gpu/gpu_
device.cc:1356] Found device 0 with properties: **
**name: Quadro M6000 24GB major: 5 minor: 2 memoryClockRate(GHz): 1.114**
**pciBusID: 0000:04:00.0**
**totalMemory: 23.89GiB freeMemory: 23.29GiB**
**2017-05-20  06:54:34.460600:  I  tensorflow/core/common_runtime/gpu/gpu_
device.cc:1435] Adding visible gpu devices: 0**
**2017-05-20  06:54:34.708584:  I  tensorflow/core/common_runtime/gpu/gpu_
device.cc:923] Device interconnect StreamExecutor with strength 1 edge matrix:**
**2017-05-20  06:54:34.708635:  I  tensorflow/core/common_runtime/gpu/gpu_
device.cc:929]      0 **
**2017-05-20  06:54:34.708644:  I  tensorflow/core/common_runtime/gpu/gpu_
device.cc:942] 0:   N **
**2017-05-20  06:54:34.709069:  I  tensorflow/core/common_runtime/gpu/gpu_
device.cc:1053]  Created  TensorFlow  device  (/job:localhost/replica:0/task:0/
device:GPU:0 with 22598 MB memory) -> physical GPU (device: 0, name**: Quadro
M6000 24GB, pci bus id: 0000:04:00.0, compute capability: 5.2)**
>>> print(sess.run(hello))
hello
```

上面的代码明确地显示，开发者正在使用 GPU！如果看不到这段代码，说明开发者并没有成功安装 GPU，而是在使用 CPU。

2.6 文档

在本章的最后，笔者想分享一下文档的重要性。很多工程师不够重视文档，只有在软件发布时才会仓促写一些文档，如果公司的管理层不重视，文档就会流于空谈。文档的作用是为了交流，便于团队间的沟通学习，用于未来的你和现在的你交流，所以文档要清楚、简洁。

这里笔者推荐一下 Markdown 文件格式，比较权威的定义是，Markdown 是一种轻量级标记语言，创始人为约翰·格鲁伯（John Gruber）。它允许人们“使用易读易写的纯文本格式编写文档”。它非常轻，普通的文本编辑器就可以使用，可随时随地写作，不用额外购买其他软件。另外，它的语法非常简单，常用的语法用 15 分钟就可以记住。

Markdown 的表现形式也非常丰富，成文后的效果和 Word、PDF 没有太大的区别。这个工具非常好用，具有丰富的表现形式，而且可使用通用管理工具进行管理，比如通过 git 就可以实现多人同时共享工作的目的。本书就是用 Markdown 写成的，可转成 Word 或 PDF 文档，从第一次执行 git check in 命令，大概经历了 200 多次入仓，才算基本成书。

3

第 3 章 基于移动端的机器学习的开发方式和流程

在介绍各种机器学习的开发之前，让我们先大概介绍一下基于移动设备机器学习的开发方式和流程。考虑到移动设备的特殊性，基于移动设备的机器学习与传统的基于网页、云计算和数据中心的机器学习的方法有很大不同。以往的系统很多都是相对静态和可控的，而且理想状态下的计算资源是无限的，团队可以更专注于算法和系统的提高。

3.1 开发方式和流程简介

现有的开发和应用大概可以分为以下四种。

第一种，在本地或云端进行机器学习的训练，训练后的模型直接运行在移动设备上。在这种情况下，云端和设备是完全分离的。在设备上进行推理和运算。

第二种，和第一种相似，但是会以云端或其他方式对在设备上的机器学习进行不断地更新和修正。在这种方式下，移动端也是以推理为主的。

第三种，即完全的云端和移动端设备交互的机器学习方式。云端对正在设备上进行的机器学习进行监测和修改，把修改过的模型推送到移动端，进行新的机器学习的推测。同时，设备会把新的机器学习的结果推送到云端。在云端，做进一步地训练，把新的增量结果推送到设备上。这样，云端和设备端就形成了一个完整的闭环。在这种方式下，基本还是延续了云端作为机器学习训练的主要资源，设备端作为机器学习推理的主要资源。

第四种，较前三种更为复杂。它考虑到机器设备数量众多，所以同时进行机器学习的训练和推理，并把机器学习训练和推理的结果推送到云端，在云端进行进一步的训练，并把训练后的新的模型推送回设备端。

第一种方式可能更适用于应用的原型的开发，或者新模型开发的预研阶段；第二、三种方式比较常见，已经大规模地应用在开发中；第四种比较复杂，需要云端和设备端的高度配合和比较复杂的算法设计，这种应用比较少。图 3-1 展示了几种机器学习方式。

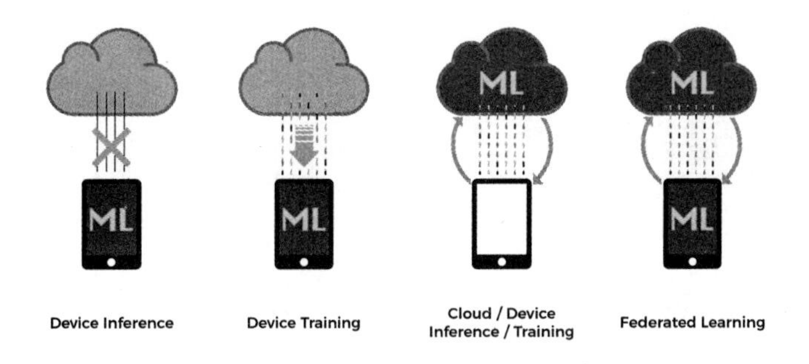

图 3-1　机器学习方式

典型的基于移动端的开发流程如下：

- 构建云端或本地的模型和测试。
- 移动端的运行、测试和改进。
- 移动端产品化及用户的使用。

由于技术资源和设计规模的限制，移动端的机器学习开发仍然严重依赖于移动设备以外的开发。它与移动应用程序的开发非常相似，通常在桌面机上设计和构建应用程序，然后在移动设备上运行应用程序。我们还要将来自移动设备的数据与应用程序连接起来，并迭代回原始设计。然后，启动另一个循环，以进一步改进应用程序。这个流程如图 3-2 所示。

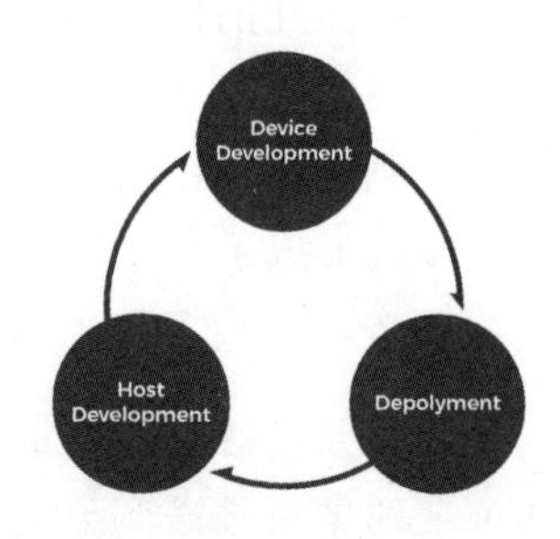

图 3-2　移动端的机器学习开发流程

对于设备上的机器学习，我们还要在主机设备上开发原始模型，对其进行评估，并将其集成到移动应用程序中。我们从运行结果和日志中收集信息，并与期望值进行比较，再循环到下一个开发周期。

3.2　使用 TPU 进行训练

现在大部分人工智能训练，都是在云端或者本地机上进行的。在本地机上用 GPU 进行训练是比较普遍的方式，在此不再做过多介绍。下面介绍一下谷歌用 TPU 进行机器学习训练的方法。

首先开发者要注册一个谷歌账号，这个账号不一定是 Gmail 的，账号也可以用各个公司的邮件地址。有了谷歌账号以后，就可以在谷歌云进行机器学习的训练了。注意，谷歌云和 TPU 的使用现在都不是免费的。

首先，用谷歌账号登录谷歌云，新建一个工程，这里新建了一个名为 tpuml 的工程，如图 3-3 所示。

然后，激活机器学习 API 和 TPU 的 API。如图 3-4 所示是激活机器学习 API 的页面；

如图 3-5 所示是激活 TPU 的 API 页面，可以看到使用时间和费用。

Google Cloud Platform

New Project

You have 24 projects remaining in your quota. Request an increase or delete projects.
Learn more

MANAGE QUOTAS

Project Name *
tpuml

Project ID: tpuml-228101. It cannot be changed later.　EDIT

Location *
No organization　BROWSE
Parent organization or folder

CREATE　CANCEL

图 3-3　工程页面

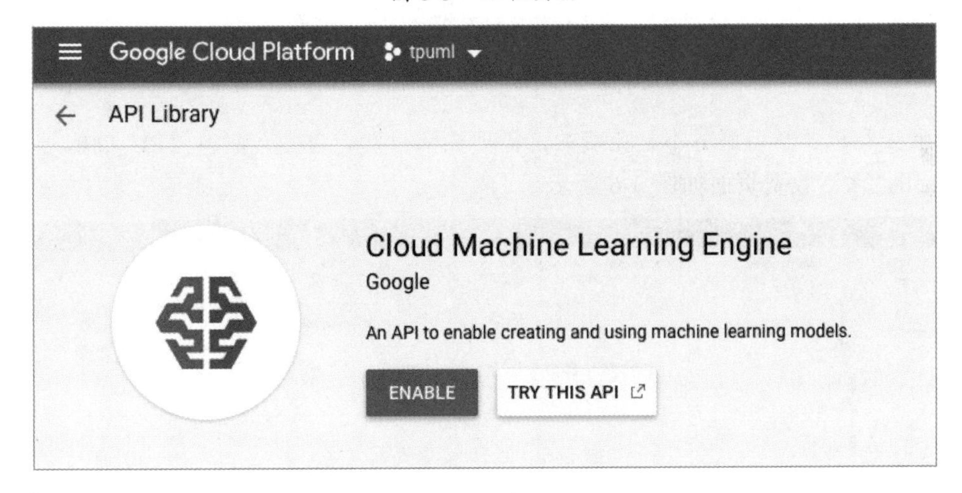

图 3-4　激活机器学习 API 的页面

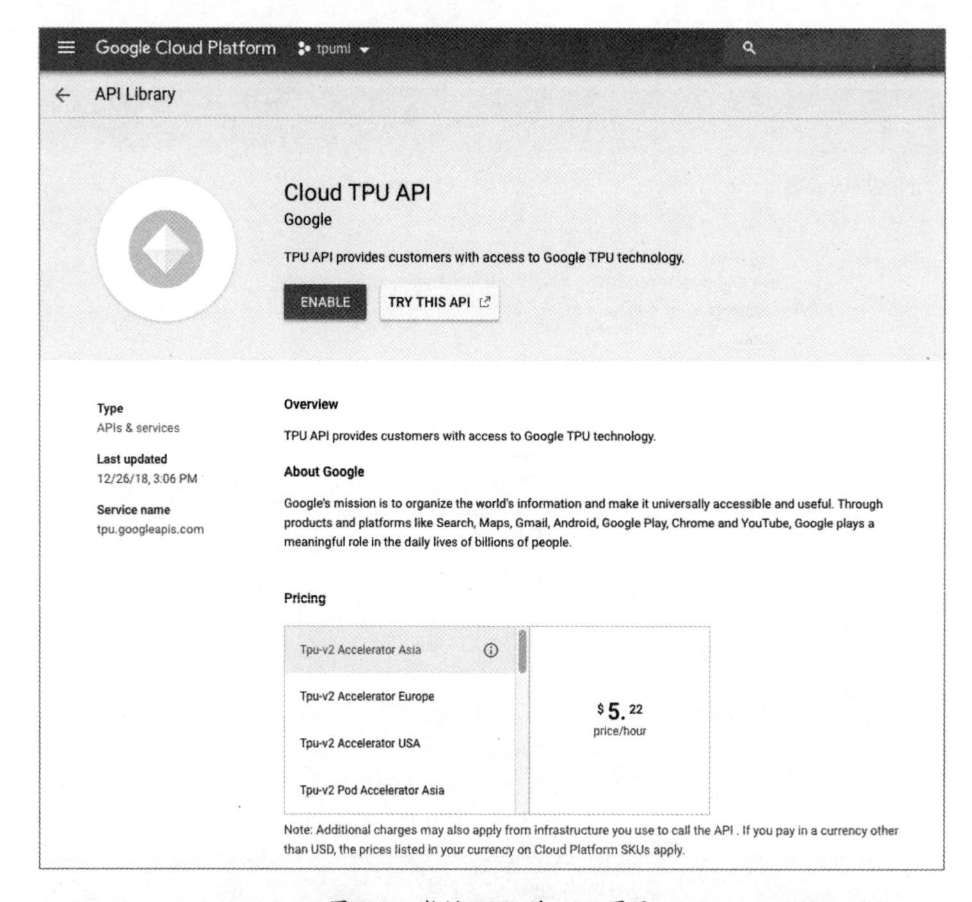

图 3-5　激活 TPU 的 API 页面

谷歌云和机器学习服务不是免费的，需要激活付费。首先，取得 TPU 的账号，然后启动 Shell，激活收费页面如图 3-6 所示。

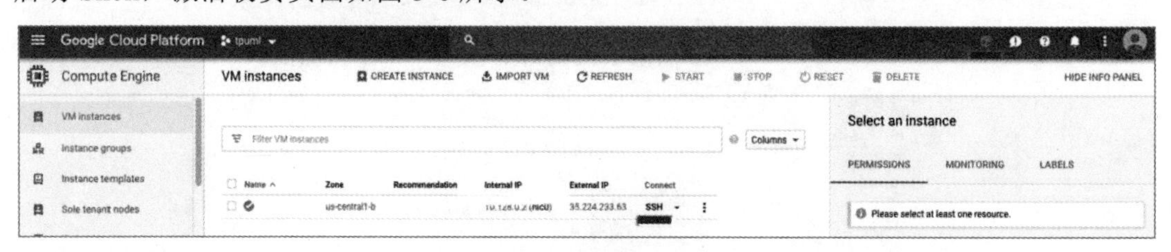

图 3-6　激活收费页面

项目在建立时会生成一个虚拟机，如图 3-6 所示，蓝色部分是虚拟机相对应的 Shell。同时，每个项目又会分配一个 Docker 影像，如图 3-6 所示，红色的图标可以启动相应的

Shell。启动 Shell 后，使用下面的命令，可以得到 TPU 的账号。

```
    $ curl -H "Authorization: Bearer $(gcloud auth print-access-token)"
https://ml.googleapis.com/v1/projects/${PROJECT}:getConfig
    /google/data/ro/teams/cloud-sdk/gcloud/google3/third_party/py/cryptograp
hy/hazmat/primitives/constant_time.py:26:    CryptographyDeprecationWarning:
Support for your Python version is deprecated. The next version of cryptography
will remove support. Please upgrade to a 2.7.x release that supports hmac.
compare_digest as soon as possible.
    {
      "serviceAccount":       "service-1049093715907@cloud-ml.google.com.iam.
gserviceaccount.com",
      "serviceAccountProject": "557869968812",
      "config": {
        "tpuServiceAccount": "service-557869968812@cloud-tpu.iam.
gserviceaccount.com"
      }
    }
```

在获取账号后，把它加到“收费账户”里。如图 3-7 所示是激活收费账号的页面。这里提醒一下读者，如果你是个人开发者，要确认 TPU 的收费方式，确保收到的账单是正确的。

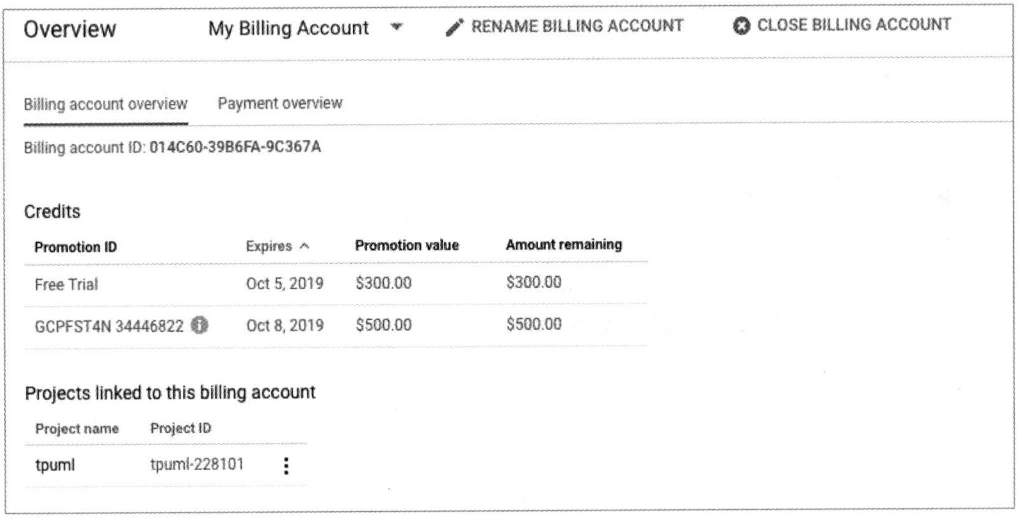

图 3-7　激活收费账号的页面

这时，就可以使用 TPU 进行机器学习的训练了。

首先，打印当前 ctpu 的设置状况，命令如下：

```
$ ctpu print-config
ctpu configuration:
        name: project_name
        project: tpuml-228101
        zone: us-central1-b
If you would like to change the configuration for a single command invocation,
please use the command line flags.
```

然后，执行 ctpu up 命令，启动 TPU。

```
cloudshell:~ (tpuml-228221)$ ctpu up
ctpu will use the following configuration:

  Name:                  project_name
  Zone:                  us-central1-b
  GCP Project:            tpuml-228221
  TensorFlow Version:     1.12
  VM:
      Machine Type:      n1-standard-2
      Disk Size:         250 GB
      Preemptible:       false
  Cloud TPU:
      Size:              v2-8
      Preemptible:       false

OK to create your Cloud TPU resources with the above configuration? [Yn]: y
```

当ctpu up命令执行完毕后，验证Shell提示符已从username@project更改为username@tpuname，这表明当前已登录计算引擎 VM 中。

接下来，建一个叫作 tpuml-bucket 的存储区域，如图 3-8 所示。

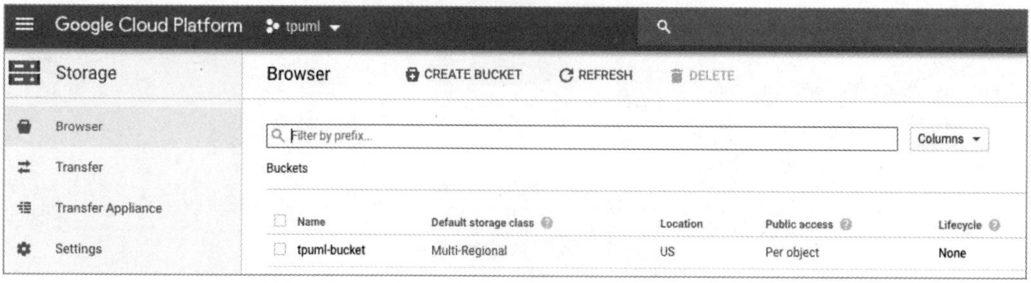

Google Cloud Platform   tpuml

Storage | Browser | CREATE BUCKET | REFRESH | DELETE

Browser
Transfer
Transfer Appliance
Settings

Filter by prefix...   Columns

Buckets

Name	Default storage class	Location	Public access	Lifecycle
tpuml-bucket	Multi-Regional	US	Per object	None

图 3-8　存储页面

在这里，我们直接使用 TPU 虚拟机自带的程序进行训练。

首先，设定 GCS_BUCKET_NAME：

```
export GCS_BUCKET_NAME=tpuml-bucket
```

然后，执行下面的命令，就可以进行 MNIST 的训练。MNIST 是手写数字的数据集，使用这个数据集进行训练，机器学习系统可以分辨出手写的数字。

```
    $ python /usr/share/tensorflow/tensorflow/examples/how_tos/reading_data/convert_to_records.py --directory=./data
    $ gunzip ./data/*.gz
    $ gsutil cp -r ./data gs://$GCS_BUCKET_NAME/mnist/data
    $  python  /usr/share/models/official/mnist/mnist_tpu.py  --data_dir=gs://$GCS_BUCKET_NAME/mnist/data/ --model_dir=gs://$GCS_BUCKET_NAME/mnist/model --tpu=$TPU_NAME
```

执行下面的命令，可以训练 ResNet-50。

```
    $ python \
        /usr/share/tpu/models/official/resnet/resnet_main.py \
        --data_dir=gs://cloud-tpu-test-datasets/fake_imagenet \
        --model_dir=gs://$GCS_BUCKET_NAME/resnet \
        --tpu=$TPU_NAME
```

ResNet-50 训练的中间结果如图 3-9 所示。

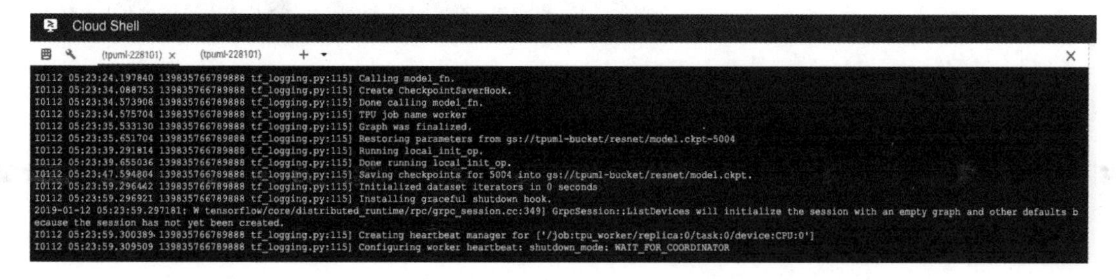

图 3-9　ResNet-50 训练的中间结果

在训练的过程中，可以通过另外一个 Shell 启动 TensorBoard：

```
$ tensorboard -logdir gs://$GCS_BUCKET_NAME/resnet &
```

TensorBoard 训练的中间结果如图 3-10 所示。

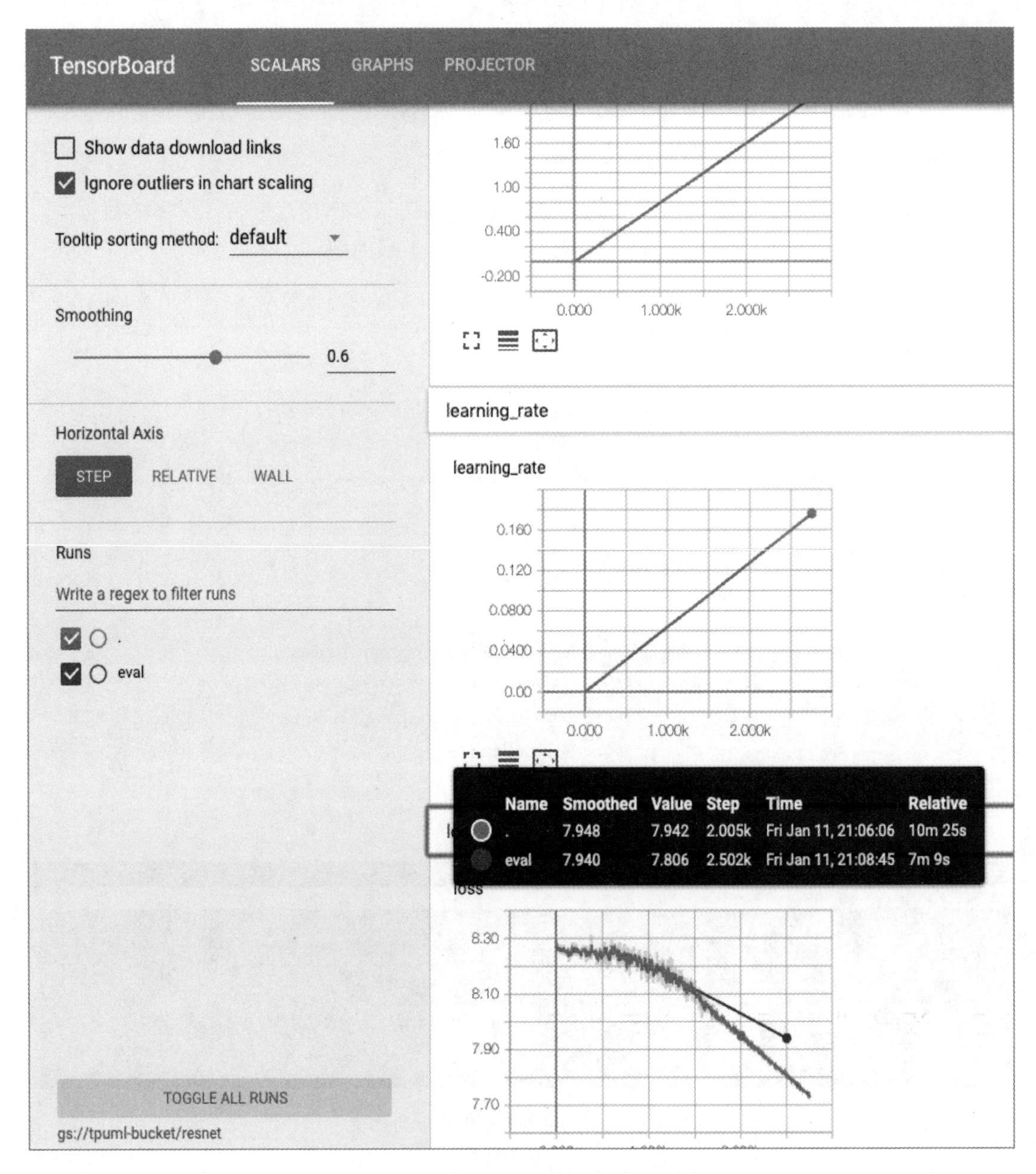

图 3-10 TensorBoard 训练的中间结果

TensorBoard 训练的中间结果会被存储在刚才建立的 Bucket 里，如图 3-11 所示显示了存储训练的结果。

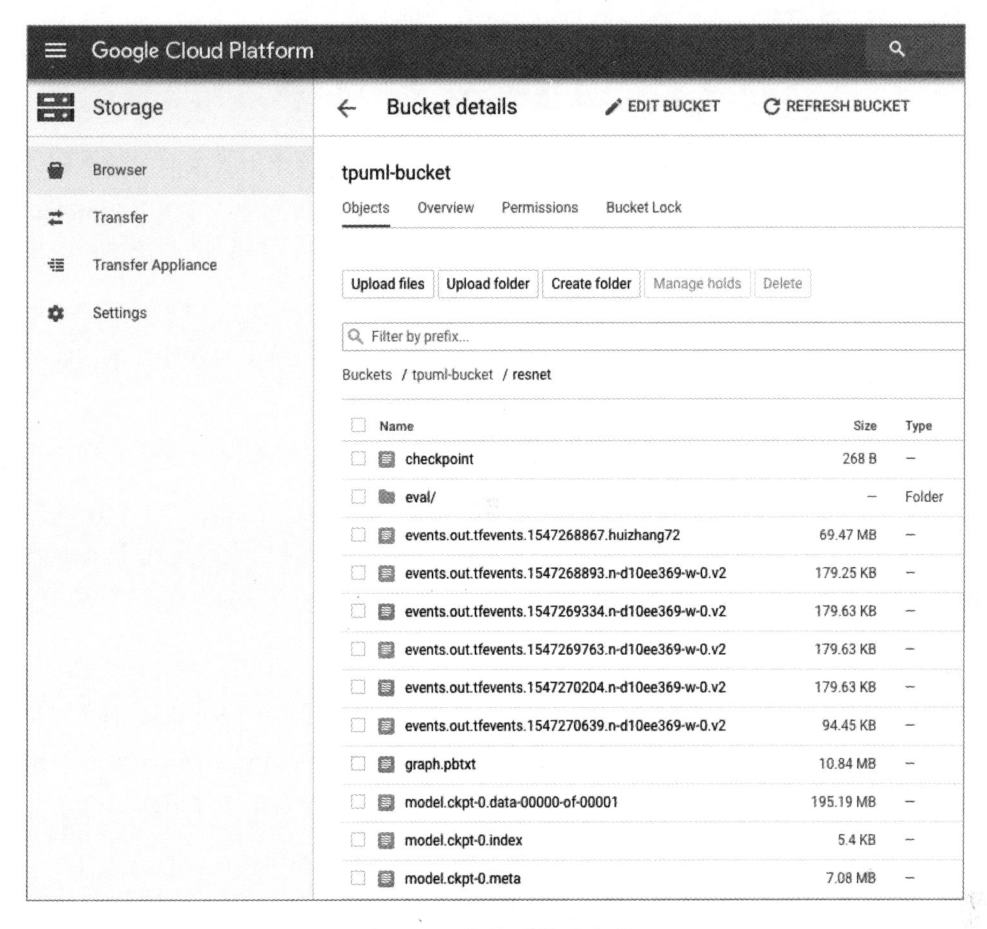

图 3-11　存储训练的结果

3.3　设备端进行机器学习训练

前面我们学习了使用谷歌云和 TPU 进行机器学习的训练。谷歌云提供了一个比较完善的机器学习开发环境，用户体验和本地机没有什么差别。

下面我们将使用谷歌开源的物体识别模型，训练出一个可以用在移动端的模型。模型的源代码地址为 https://github.com/tensorflow/models/blob/master/research/object_detection。

我们继续使用上面建立的 tpuml 项目，请留意下面三个名称的不同意义：

```
Project name: tpuml
Project ID: tpuml-22801
Project number: 10909371907
```

project name 是工程名称（可以不唯一），project ID 是程序中真正的工程名称（唯一的），project number 是唯一的工程数字编号。我们以前提到了谷歌经常使用 Protobuf 作为数据定义的格式，估计在这里，一个域（Field）是名称，一个域是 ID，这也是常用的定义方式。

另外，为这次训练，我们新建了一个叫作 tpuml-bucket/object 的存储区域，并且设置相应的环境变量，代码如下：

```
$ export YOUR_GCS_BUCKET=tpuml-bucket/object
```

这次我们将要训练一个能识别宠物的机器学习模型，选用 Oxford-IIIT Pets 数据集。首先，要把数据下载下来；然后，转换成 TensorFlow 可以使用的数据格式 TFRecord；最后，上传到刚才建立的云存储里。

接下来介绍一下 TFRecord。

TFRecord 是一种 TensorFlow 对应的文件格式，它是通过 tf_record.py 脚本（https://github.com/ tensorflow/tensorflow/blob/master/tensorflow/python/lib/io/tf_record.py）来实现的。TFRecord 可以支持压缩格式，是 TensorFlow 通用的数据存储文件格式。

由于接下来要用到 object_detection，所以我们要先下载源代码：

```
$ git clone https://github.com/tensorflow/models.git
$ cd models/research/
```

然后，下载数据和处理数据：

```
# 下载数据集
$ wget http://www.robots.ox.ac.uk/~vgg/data/pets/data/images.tar.gz
$ wget http://www.robots.ox.ac.uk/~vgg/data/pets/data/annotations.tar.gz

# 解压下载的数据集
$ tar -xvf images.tar.gz
$ tar -xvf annotations.tar.gz

# 在models/research 里
$ python object_detection/dataset_tools/create_pet_tf_record.py \
   --label_map_path=object_detection/data/pet_label_map.pbtxt \
```

```
        --data_dir=`pwd` \
        --output_dir=`pwd`

    # 在 tensorflow/models/research 里
    $ gsutil cp pet_faces_train.record-* gs://${YOUR_GCS_BUCKET}/data/
    $ gsutil cp pet_faces_val.record-* gs://${YOUR_GCS_BUCKET}/data/
    $    gsutil    cp    object_detection/data/pet_label_map.pbtxt    gs:
//${YOUR_GCS_BUCKET}/data/pet_label_map.pbtxt
```

下面，我们下载 COCO 预训练模型进行迁移学习（Transfer Learning）。如果从头开始训练物体探测器，那么会需要大量的时间。为了加速训练，我们将采用在不同数据集（COCO）上训练的物体探测器，并重复使用其中的一些参数来初始化新模型。

```
    # 下载模型
    $   wget   http://storage.googleapis.com/download.tensorflow.org/models/
object_detection/faster_rcnn_resnet101_coco_11_06_2017.tar.gz
    $ tar -xvf faster_rcnn_resnet101_coco_11_06_2017.tar.gz
    $    gsutil    cp    faster_rcnn_resnet101_coco_11_06_2017/model.ckpt.*
gs://${YOUR_GCS_BUCKET}/data/

    # 把存储的路径转换到设定文件中
    $ sed -i "s|PATH_TO_BE_CONFIGURED|"gs://${YOUR_GCS_BUCKET}"/data|g" \
        object_detection/samples/configs/faster_rcnn_resnet101_pets.config

    # 将设定文件复制到谷歌存储中
    $   gsutil   cp   object_detection/samples/configs/faster_rcnn_resnet101_
pets.config \
        gs://${YOUR_GCS_BUCKET}/data/faster_rcnn_resnet101_pets.config
```

至此，存储区域里应该有以下中间结果文件，如图 3-12 所示。

接着，把刚才准备的文件打包，执行下面的命令：

```
    $ bash object_detection/dataset_tools/create_pycocotools_package.sh /tmp
/pycocotools
    $ python setup.py sdist
    $ (cd slim && python setup.py sdist)
```

最后，可以执行训练了。在这里，训练和评估（Evaluation）可以同时进行，代码如下：

```
    $   gcloud   ml-engine   jobs   submit   training   `whoami`_object_detection_
pets_`date +%m_%d_%Y_%H_%M_%S` \
```

```
        --runtime-version 1.9 \
        --job-dir=gs://${YOUR_GCS_BUCKET}/model_dir \
        --packages dist/object_detection-0.1.tar.gz,slim/dist/slim-0.1.tar.gz,
/tmp/pycocotools/pycocotools-2.0.tar.gz \
        --module-name object_detection.model_main \
        --region us-central1 \
        --config object_detection/samples/cloud/cloud.yml \
        -- \
        --model_dir=gs://${YOUR_GCS_BUCKET}/model_dir \
        --pipeline_config_path=gs://${YOUR_GCS_BUCKET}/data/faster_rcnn_
resnet101_pets.config
```

Bucket details   EDIT BUCKET   REFRESH BUCKET

tpuml-bucket

Objects   Overview   Permissions   Bucket Lock

Upload files   Upload folder   Create folder   Manage holds   Delete

Filter by prefix...

Buckets / tpuml-bucket / object / data

Name	Size	Type
faster_rcnn_resnet101_pets.config	3.65 KB	application/octet-stream
model.ckpt.data-00000-of-00001	425.16 MB	application/octet-stream
model.ckpt.index	39.57 KB	application/octet-stream
model.ckpt.meta	10.66 MB	application/octet-stream

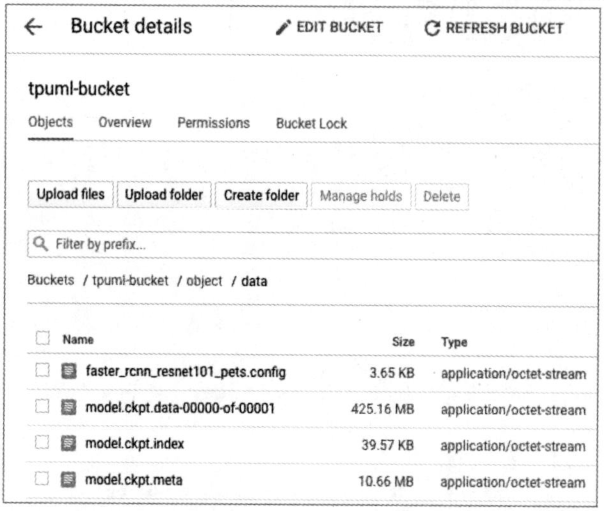

图 3-12　中间结果文件

如图 3-13 所示显示了训练中间结果。

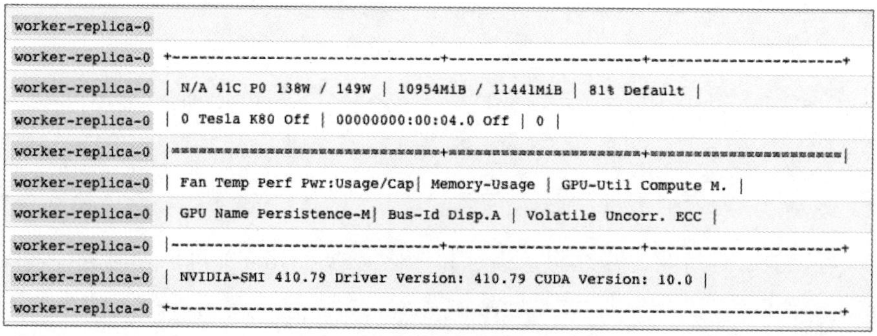

```
worker-replica-0
worker-replica-0  +-----------------------------+----------------------+----------------------+
worker-replica-0  | N/A 41C P0 138W / 149W | 10954MiB / 11441MiB | 81% Default |
worker-replica-0  | 0 Tesla K80 Off | 00000000:00:04.0 Off | 0 |
worker-replica-0  |=============================+======================+======================|
worker-replica-0  | Fan Temp Perf Pwr:Usage/Cap| Memory-Usage | GPU-Util Compute M. |
worker-replica-0  | GPU Name Persistence-M| Bus-Id Disp.A | Volatile Uncorr. ECC |
worker-replica-0  |-----------------------------+----------------------+----------------------+
worker-replica-0  | NVIDIA-SMI 410.79 Driver Version: 410.79 CUDA Version: 10.0 |
worker-replica-0  +-----------------------------------------------------------------------------+
```

图 3-13　训练中间结果

在训练的同时，我们可以打开另外一个 Shell，用 TensorBoard 去检查训练的进展即中间结果，如图 3-14 所示。

图 3-14　检查中间结果

现在就可以使用 TPU 进行训练了，代码如下：

```
# 重新输出到一个新的文件夹中
$ MODEL_DIR=tpuml-bucket/object/model2

# 训练
$ gcloud ml-engine jobs submit training `whoami`_object_detection_`date
+%m_%d_%Y_%H_%M_%S` \
  --job-dir=gs://${MODEL_DIR} \
  --packages dist/object_detection-0.1.tar.gz,slim/dist/slim-0.1.tar.gz,
```

```
/tmp/pycocotools/pycocotools-2.0.tar.gz \
      --module-name object_detection.model_tpu_main \
      --runtime-version 1.9 \
      --scale-tier BASIC_TPU \
      --region us-central1 \
      -- \
      --tpu_zone us-central1 \
      --model_dir=gs://${MODEL_DIR} \
      --pipeline_config_path=gs://${YOUR_GCS_BUCKET}/data/faster_rcnn_
resnet101_pets_tpu.config

    # 评价
    $ gcloud ml-engine jobs submit training object_detection_eval_`date
+%m_%d_%Y_%H_%M_%S` \
      --runtime-version 1.9 \
      --job-dir=gs://${MODEL_DIR} \
      --packages dist/object_detection-0.1.tar.gz,slim/dist/slim-0.1.tar.gz,
/tmp/pycocotools/pycocotools-2.0.tar.gz \
      --module-name object_detection.model_main \
      --region us-central1 \
      --scale-tier BASIC_GPU \
      -- \
      --model_dir=gs://${MODEL_DIR} \
      --pipeline_config_path=gs://${YOUR_GCS_BUCKET}/data/faster_rcnn_
resnet101_pets_tpu.config \
      --checkpoint_dir=gs://${MODEL_DIR}
```

TPU 在谷歌内部已经被广泛使用，但作为开发者，笔者在使用谷歌云 TPU 的时候，还是遇到了一些问题，主要是版本兼容和一些文档没有及时更新的问题。在实际应用的时候，可能要做一些变通，开发者一般通过公开的渠道，可以从谷歌获取答案。

在模型生成之后，我们需要将其转换成 TensorFlow Lite 能使用的模式。首先，生成 TensorFlow Lite 的模型文件，让我们先执行下面的代码：

```
    $ export CONFIG_FILE=gs://tpuml-bucket/object/data/pipeline.config
    $ export CHECKPOINT_PATH=gs://tpuml-bucket/object/data/model.ckpt
    $ export OUTPUT_DIR=/tmp/tflite

    $ object_detection/export_tflite_ssd_graph.py \
      --pipeline_config_path=$CONFIG_FILE \
      --trained_checkpoint_prefix=$CHECKPOINT_PATH \
      --output_directory=$OUTPUT_DIR \
```

```
    --add_postprocessing_op=true
```

执行后，生成如下内容：

```
tflite_graph.pb
tflite_graph.pbtxt
```

然后，执行如下代码：

```
$ bazel run --config=opt tensorflow/lite/toco:toco -- \
    --input_file=$OUTPUT_DIR/tflite_graph.pb \
    --output_file=$OUTPUT_DIR/detect.tflite \
    --input_shapes=1,300,300,3 \
    --input_arrays=normalized_input_image_tensor \
    --output_arrays='TFLite_Detection_PostProcess','TFLite_Detection_
PostProcess:1','TFLite_Detection_PostProcess:2','TFLite_Detection_PostProces
s:3' \
    --inference_type=QUANTIZED_UINT8 \
    --mean_values=128 \
    --std_values=128 \
    --change_concat_input_ranges=false \
    --allow_custom_ops
```

上面的代码执行完毕，就会得到detect.tflite文件，我们可以把这个模型文件pet_label_map.pbtxt 集成进 TensorFlow Lite 的应用，再运行在终端设备上，具体实现后面会详细介绍。

关于 TensorFlow Lite 的模型格式和转换方法，后面也会详细讲解，在这里只是大概介绍一下。

3.4　使用 TensorFlow Serving 优化 TensorFlow 模型

下面笔者来介绍一下如何使用 TensorFlow Serving 组件导出已经过训练的 TensorFlow 模型，并使用标准 tensorflow_model_server 来对它提供服务。如果想了解更多信息，请参阅 TensorFlow Serving 高阶教程（https://tensorflow.google.cn/serving/serving_advanced?hl=zh-CN）。

本书使用 TensorFlow 教程中引入的简单 Softmax 回归模型进行手写图像（MNIST 数据）分类。如果开发者不熟悉 TensorFlow 或 MNIST 是什么，可以参阅 MNIST For ML

Beginners 教程（http://tensorflow.google.cn/tutorials/mnist/beginners/index.html?hl=zh-CN #mnist-for-ml-beginners）。

本书的代码由两部分组成：

一个是 Python 文件 mnist_saved_model.py（https://github.com/tensorflow/serving/tree/master/ tensorflow_serving/example/mnist_saved_model.py），用于训练和导出模型。

另一个是 ModelServer 二进制文件，可以使用 Apt 安装，也可以从 C++文件（main.cc）编译（https://github.com/tensorflow/serving/tree/master/tensorflow_serving/model_servers/main.cc）。TensorFlow Serving 的 ModelServer 会发现新的导出模型，并运行 gRPC 服务来为其服务（http://www.grpc.io/）。gRPC 是谷歌开源的高效轻量级进程通信协议，谷歌的对外接口基本都支持这个协议，因此被很多国内互联网公司所采用。

TensorFlow 模型训练开始之前，需要安装 Docker（https://tensorflow.google.cn/serving/docker?hl=zh-CN#installing-docker）。

3.4.1 训练和导出 TensorFlow 模型

正如在 mnist_saved_model.py 中所见，训练的方式与在初学者 MNIST 教程中完成的方式相同（https://tensorflow.google.cn/get_started/mnist/beginners?hl=zh-CN）。TensorFlow 运行图在 TensorFlow 会话中启动，输入张量（图像）为 $\boldsymbol{x}$，输出张量（Softmax 得分）为 $\boldsymbol{y}$。

然后，我们使用 TensorFlow 的 SavedModelBuilder 模块导出模型，代码如下：

```
"""SavedModel builder.
Builds a SavedModel that can be saved to storage, is language neutral, and
enables systems to produce, consume, or transform TensorFlow Models.
"""

from __future__ import absolute_import
from __future__ import division
from __future__ import print_function

from tensorflow.python.saved_model.builder_impl import _SavedModelBuilder
from tensorflow.python.saved_model.builder_impl import SavedModelBuilder
```

SavedModelBuilder 将训练模型的“快照”保存到可靠存储中，以便稍后加载并进行推理。有关 SavedModel 格式的详细信息，请参阅 SavedModel 中的 README.md 文档

（https://github.com/tensorflow/tensorflow/blob/master/tensorflow/python/saved_model/READ-M E.md）。

下面的代码（https://github.com/tensorflow/serving/tree/master/tensorflow_serving/example/mnist_saved_model.py）可以训练和导出 minst 模型：

```
#! /usr/bin/env python
r"""Train and export a simple Softmax Regression TensorFlow model.
The model is from the TensorFlow "MNIST For ML Beginner" tutorial. This program
simply follows all its training instructions, and uses TensorFlow SavedModel to
export the trained model with proper signatures that can be loaded by standard
tensorflow_model_server.
Usage: mnist_saved_model.py [--training_iteration=x] [--model_version=y] \
    export_dir
"""

from __future__ import print_function

import os
import sys

import tensorflow as tf

import mnist_input_data

tf.app.flags.DEFINE_integer('training_iteration', 1000,
                            'number of training iterations.')
tf.app.flags.DEFINE_integer('model_version', 1, 'version number of the
model.')
tf.app.flags.DEFINE_string('work_dir', '/tmp', 'Working directory.')
FLAGS = tf.app.flags.FLAGS


def main(_):
  if len(sys.argv) < 2 or sys.argv[-1].startswith('-'):
    print('Usage: mnist_saved_model.py [--training_iteration=x] '
          '[--model_version=y] export_dir')
    sys.exit(-1)
  if FLAGS.training_iteration <= 0:
    print('Please specify a positive value for training iteration.')
    sys.exit(-1)
```

```
    if FLAGS.model_version <= 0:
      print('Please specify a positive value for version number.')
      sys.exit(-1)

    # 训练模型
    print('Training model...')
    mnist = mnist_input_data.read_data_sets(FLAGS.work_dir, one_hot=True)
    sess = tf.InteractiveSession()
    serialized_tf_example = tf.placeholder(tf.string, name='tf_example')
    feature_configs   =   {'x':   tf.FixedLenFeature(shape=[784],   dtype=
tf.float32),}
    tf_example = tf.parse_example(serialized_tf_example, feature_configs)
    x = tf.identity(tf_example['x'], name='x')  # use tf.identity() to assign
name
    y_ = tf.placeholder('float', shape=[None, 10])
    w = tf.Variable(tf.zeros([784, 10]))
    b = tf.Variable(tf.zeros([10]))
    sess.run(tf.global_variables_initializer())
    y = tf.nn.softmax(tf.matmul(x, w) + b, name='y')
    cross_entropy = -tf.reduce_sum(y_ * tf.log(y))
    train_step  =  tf.train.GradientDescentOptimizer(0.01).minimize(cross_
entropy)
    values, indices = tf.nn.top_k(y, 10)
    table = tf.contrib.lookup.index_to_string_table_from_tensor(
        tf.constant([str(i) for i in range(10)]))
    prediction_classes = table.lookup(tf.to_int64(indices))
    for _ in range(FLAGS.training_iteration):
      batch = mnist.train.next_batch(50)
      train_step.run(feed_dict={x: batch[0], y_: batch[1]})
    correct_prediction = tf.equal(tf.argmax(y, 1), tf.argmax(y_, 1))
    accuracy = tf.reduce_mean(tf.cast(correct_prediction, 'float'))
    print('training accuracy %g' % sess.run(
        accuracy, feed_dict={
            x: mnist.test.images,
            y_: mnist.test.labels
        }))
    print('Done training!')

    # 导出模型
    export_path_base = sys.argv[-1]
    export_path = os.path.join(
```

```
        tf.compat.as_bytes(export_path_base),
        tf.compat.as_bytes(str(FLAGS.model_version)))
    print('Exporting trained model to', export_path)
    builder = tf.saved_model.builder.SavedModelBuilder(export_path)

    # 创建 signature_def_map
    classification_inputs = tf.saved_model.utils.build_tensor_info(
        serialized_tf_example)
    classification_outputs_classes = tf.saved_model.utils.build_
tensor_info(
        prediction_classes)
    classification_outputs_scores = tf.saved_model.utils.build_tensor_
info(values)

    classification_signature = (
        tf.saved_model.signature_def_utils.build_signature_def(
            inputs={
                tf.saved_model.signature_constants.CLASSIFY_INPUTS:
                    classification_inputs
            },
            outputs={
                tf.saved_model.signature_constants.CLASSIFY_OUTPUT_CLASSES:
                    classification_outputs_classes,
                tf.saved_model.signature_constants.CLASSIFY_OUTPUT_SCORES:
                    classification_outputs_scores
            },
            method_name=tf.saved_model.signature_constants.CLASSIFY_METHOD_
NAME))

    tensor_info_x = tf.saved_model.utils.build_tensor_info(x)
    tensor_info_y = tf.saved_model.utils.build_tensor_info(y)

    prediction_signature = (
        tf.saved_model.signature_def_utils.build_signature_def(
            inputs={'images': tensor_info_x},
            outputs={'scores': tensor_info_y},
            method_name=tf.saved_model.signature_constants.PREDICT_METHOD_
NAME))

    builder.add_meta_graph_and_variables(
        sess, [tf.saved_model.tag_constants.SERVING],
```

```
        signature_def_map={
            'predict_images':
                prediction_signature,
            tf.saved_model.signature_constants.DEFAULT_SERVING_SIGNATURE_
DEF_KEY:
                classification_signature,
        },
        main_op=tf.tables_initializer(),
        strip_default_attrs=True)

    builder.save()

    print('Done exporting!')


if __name__ == '__main__':
    tf.app.run()
```

下面这个简短的通用代码片段的功能是将模型保存到机器磁盘。

```
export_path_base = sys.argv[-1]

export_path = os.path.join(
    compat.as_bytes(export_path_base),
    compat.as_bytes(str(FLAGS.model_version)))
print 'Exporting trained model to', export_path
builder = tf.saved_model.builder.SavedModelBuilder(export_path)
builder.add_meta_graph_and_variables(
    sess, [tag_constants.SERVING],
    signature_def_map={
       'predict_images':
           prediction_signature,
       signature_constants.DEFAULT_SERVING_SIGNATURE_DEF_KEY:
           classification_signature,
    },
    main_op=main_op)

    builder.save()
```

SavedModelBuilder 采用以下参数：

export_path 导出目录的路径。如果目录不存在，SavedModelBuilder 将创建该目录。在示例中，我们连接命令行参数和 FLAGS.model_version 以获取导出目录。

FLAGS.model_version 指定模型的版本。在导出同一模型的较新版本时，应指定较大的整数值。每个版本将导出到指定路径下的不同子目录中。

开发者可以通过 SavedModelBuilder.add_meta_graph_and_variables()函数将元图和变量添加到构建器中。该函数的相关参数解释如下：

sess TensorFlow 会话，包含正在导出的训练模型。

tags 用于保存元图的标记集。由于我们打算在服务中使用图形，因此我们使用来自预定义的 SavedModel 标记常量的 serve 标记。 更多详细信息，请参阅 tag_constants.py（ https://github.com/tensorflow/tensorflow/blob/master/tensorflow/python/saved_model/tag_constants.py）和相关的 TensorFlow API 文档（https://tensorflow.google.cn/api_docs/python/tf/saved_model/tag_constants?hl=zh-CN）。

signature_def_map 将指定用户提供的签名映射到 tensorflow::SignatureDef，Signature 指定正在导出的模型类型，以及在运行推理时绑定的输入/输出张量。

serving_default 特殊签名密钥，用于指定默认服务签名。默认服务签名 def 键及与签名相关的其他常量被定义为 SavedModel 签名常量的一部分。更多详细信息请参阅 signature_constants.py（https://github.com/tensorflow/tensorflow/blob/master/tensorflow/python/saved_model/ signature_constants.py）和相关的 TensorFlow 1.0 API 文档（https://tensorflow.google.cn/api_docs/ python/tf/saved_model/signature_constants?hl= zh-CN）。

此外，为了构建签名定义，SavedModel API 提供了 Module: tf.saved_model.signature_def_utils （ https://tensorflow.google.cn/api_docs/python/tf/saved_model/signature_def_utils?hl=zh-CN）具体地说，在原始的 mnist_saved_model.py 文件中（https://github.com/tensorflow/serving/tree/ master/tensorflow_serving/example/mnist_saved_model.py），我们将 signature_def_utils.build_ signature_def( )用于 buildpredict_signature 和 classification_ signature。

那么，如何使用 predict_signature 呢？相关参数解释如下：

inputs={'images': tensor_info_x}　指定输入张量信息。

outputs={'scores': tensor_info_y} 指定分数张量信息。

method_name 是用于推理的方法。对于预测请求，应将其设置为 tensorflow / serving / predict。有关其他方法名称，请参阅 signature_constants.py（ https://github.com/tensorflow/tensorflow/blob/master/tensorflow/python/saved_model/signature_constants.py）和 relatedTensorFlow

1.0 API 文档（https://tensorflow.google.cn/api_docs/python/tf/saved_model/ signature_constants?hl=zh-CN）。

请注意，tensor_info_x 和 tensor_info_y 具有此处定义的 tensorflow::TensorInfo 协议缓冲区的结构（https://github.com/tensorflow/tensorflow/blob/master/tensorflow/core/protobuf/meta_graph. proto）。

为了构建张量信息，TensorFlow SavedModel API 还提供了 utils.py，代码如下：

```
    # Copyright 2016 The TensorFlow Authors. All Rights Reserved.
    #
    # Licensed under the Apache License, Version 2.0 (the "License");
    # you may not use this file except in compliance with the License.
    # You may obtain a copy of the License at
    #
    #     http://www.apache.org/licenses/LICENSE-2.0
    #
    # Unless required by applicable law or agreed to in writing, software
    # distributed under the License is distributed on an "AS IS" BASIS,
    # WITHOUT WARRANTIES OR CONDITIONS OF ANY KIND, either express or implied.
    # See the License for the specific language governing permissions and
    # limitations under the License.
    # ==============================================================================
    """SavedModel utility functions.
    Utility functions to assist with setup and construction of the SavedModel
proto.
    """
    from __future__ import absolute_import
    from __future__ import division
    from __future__ import print_function

    # pylint: disable=unused-import
    from tensorflow.python.saved_model.utils_impl import build_tensor_info
    from tensorflow.python.saved_model.utils_impl import build_tensor_info_
from_op
    from tensorflow.python.saved_model.utils_impl import get_tensor_from_
tensor_info
    # pylint: enable=unused-import
```

源代码的地址为 https://github.com/tensorflow/tensorflow/blob/master/tensorflow/python/

saved_model/utils.py，相关文档的地址为 https://tensorflow.google.cn/api_docs/python/tf/saved_model/utils?hl=zh-CN。

另外一点需要注意的是，图像和分数是张量别名，它们可以是开发者想要的任何特定的字符串。作为张量 ***x*** 和 ***y*** 的逻辑名称，可以在稍后发送预测请求时引用张量进行绑定。

例如，如果 ***x*** 引用名称为“long_tensor_name_foo”的张量，***y*** 引用名称为“generated_tensor_name_bar”的张量，则构建器将张量逻辑名称存储为实名映射：

```
'images' -> 'long_tensor_name_foo'
'score' -> 'generated_tensor_name_bar'
```

用户在运行推理时可以使用其逻辑名称来引用这些张量。

下面就可以运行了。

首先，克隆原始代码到本地：

```
$ git clone https://github.com/tensorflow/serving.git
$ cd serving
```

然后，清除输出目录，结果会输出到这个目录中：

```
rm -rf /tmp/mnist
```

现在，让我们训练模型：

```
$ tools/run_in_docker.sh python tensorflow_serving/example/mnist_saved_model.py /tmp/mnist
```

运行过程的屏幕输出如下：

```
Training model...

...

Done training!
Exporting trained model to models/mnist
Done exporting!
```

查看导出的目录：

```
$ ls /tmp/mnist

1
```

如上所述，为了导出模型的每个版本，将会创建一个子目录。FLAGS.model_version 的默认值为 1，因此创建了相应的子目录 1，代码如下：

```
$ ls /tmp/mnist/1
saved_model.pb variables
```

每个版本的子目录下都包含 saved_model.pb 和 variables 两个文件。saved_model.pb 是序列化的 tensorflow::SavedModel，它包括模型的一个或多个图形定义，以及模型的元数据（如签名）；Variable 是包含图形序列化变量的文件。

3.4.2 使用标准 TensorFlow ModelServer 加载导出的模型

使用 Docker 服务加载模型，代码如下：

```
$ docker run -p 8500:8500 \
  --mount type=bind,source=/tmp/mnist,target=/models/mnist \
  -e MODEL_NAME=mnist -t tensorflow/serving &
```

3.4.3 测试服务器

我们可以使用下面的 mnist_client 程序（源代码在 https://github.com/tensorflow/serving/tree/master/tensorflow_serving/example/mnist_client.py）来测试服务器，该客户端程序下载 MNIST 测试数据，将其作为请求发送给服务器，并计算推理错误率。

```
#!/usr/bin/env python2.7

"""A client that talks to tensorflow_model_server loaded with mnist
model.
The client downloads test images of mnist data set, queries the service
with
such test images to get predictions, and calculates the inference error
rate.
Typical usage example:
    mnist_client.py --num_tests=100 --server=localhost:9000
"""

from __future__ import print_function

import sys
```

```
    import threading

    import grpc
    import numpy
    import tensorflow as tf

    from tensorflow_serving.apis import predict_pb2
    from tensorflow_serving.apis import prediction_service_pb2_grpc
    import mnist_input_data


    tf.app.flags.DEFINE_integer('concurrency', 1,
                                'maximum  number  of  concurrent  inference
requests')
    tf.app.flags.DEFINE_integer('num_tests', 100, 'Number of test images')
    tf.app.flags.DEFINE_string('server',      '',      'PredictionService
host:port')
    tf.app.flags.DEFINE_string('work_dir', '/tmp', 'Working directory. ')
    FLAGS = tf.app.flags.FLAGS


    class _ResultCounter(object):
      """Counter for the prediction results."""

      def __init__(self, num_tests, concurrency):
        self._num_tests = num_tests
        self._concurrency = concurrency
        self._error = 0
        self._done = 0
        self._active = 0
        self._condition = threading.Condition()

      def inc_error(self):
        with self._condition:
          self._error += 1

      def inc_done(self):
        with self._condition:
          self._done += 1
          self._condition.notify()

      def dec_active(self):
```

```
    with self._condition:
      self._active -= 1
      self._condition.notify()

  def get_error_rate(self):
    with self._condition:
      while self._done != self._num_tests:
        self._condition.wait()
      return self._error / float(self._num_tests)

  def throttle(self):
    with self._condition:
      while self._active == self._concurrency:
        self._condition.wait()
      self._active += 1
```

```
def _create_rpc_callback(label, result_counter):
  """Creates RPC callback function.
  Args:
    label: The correct label for the predicted example.
    result_counter: Counter for the prediction result.
  Returns:
    The callback function.
  """
  def _callback(result_future):
    """Callback function.
    Calculates the statistics for the prediction result.
    Args:
      result_future: Result future of the RPC.
    """
    exception = result_future.exception()
    if exception:
      result_counter.inc_error()
      print(exception)
    else:
      sys.stdout.write('.')
      sys.stdout.flush()
      response = numpy.array(
          result_future.result().outputs['scores'].float_val)
      prediction = numpy.argmax(response)
      if label != prediction:
```

```
        result_counter.inc_error()
      result_counter.inc_done()
      result_counter.dec_active()
    return _callback


  def do_inference(hostport, work_dir, concurrency, num_tests):
    """Tests PredictionService with concurrent requests.
    Args:
      hostport: Host:port address of the PredictionService.
      work_dir: The full path of working directory for test data set.
      concurrency: Maximum number of concurrent requests.
      num_tests: Number of test images to use.
    Returns:
      The classification error rate.
    Raises:
      IOError: An error occurred processing test data set.
    """
    test_data_set = mnist_input_data.read_data_sets(work_dir).test
    channel = grpc.insecure_channel(hostport)
    stub = prediction_service_pb2_grpc.PredictionServiceStub(channel)
    result_counter = _ResultCounter(num_tests, concurrency)
    for _ in range(num_tests):
      request = predict_pb2.PredictRequest()
      request.model_spec.name = 'mnist'
      request.model_spec.signature_name = 'predict_images'
      image, label = test_data_set.next_batch(1)
      request.inputs['images'].CopyFrom(
        tf.contrib.util.make_tensor_proto(image[0], shape=[1, image[0].
size]))
      result_counter.throttle()
      result_future = stub.Predict.future(request, 5.0)  # 5 seconds
      result_future.add_done_callback(
          _create_rpc_callback(label[0], result_counter))
    return result_counter.get_error_rate()


  def main(_):
    if FLAGS.num_tests > 10000:
      print('num_tests should not be greater than 10k')
      return
    if not FLAGS.server:
```

```
      print('please specify server host:port')
      return
    error_rate = do_inference(FLAGS.server, FLAGS.work_dir,
                              FLAGS.concurrency, FLAGS.num_tests)
    print('\nInference error rate: %s%%' % (error_rate * 100))


  if __name__ == '__main__':
    tf.app.run()
  $  tools/run_in_docker.sh  python  tensorflow_serving/example/mnist_
client.py \
    --num_tests=1000 --server=127.0.0.1:8500
```

代码输出结果如下：

```
  Inference error rate: 11.13%
```

对于训练好的 Softmax 模型，我们预计准确率约为 90%，前 1000 个测试图像的推理错误率为 11.13%。从这个结果，我们就可以确认服务器成功加载并运行了已训练的模型！

3.5 TensorFlow 扩展（Extended）

在机器学习里，我们非常关注模型代码，但是 TensorFlow Extended 不只是模型。那么，TensorFlow Extended 解决了哪些问题呢？

机器学习的代码其实很简单，但数据收集、配置、机器管理等事情需要投入大量的时间和精力，那么这些需要花费外围力量的工作，我们是否能够通过其他方式来实现，从而让项目快速实施落地呢？

TensorFlow 扩展（Extended）就是谷歌推出的一个能够帮助开发者解决这些问题、实现项目快速实施落地的有效工具。TensorFlow 扩展是谷歌内部广泛使用的基础架构，是机器学习平台的一个主要组成部分。但是，现在还有很多代码和功能没有开源。这个架构依托于谷歌强大的基础设施，也是谷歌强大机器学习能力的一部分。

4

第 4 章 构建 TensorFlow Mobile

本章主要介绍如何构建 TensorFlow Mobile 的结构，以及如何构建应用。

4.1 TensorFlow Mobile 的历史

TensorFlow Mobile 是 TensorFlow 的第一个对移动端嵌入式设备的支持框架。在早期的 TensorFlow 发布中就加入了对包括 Android、iOS 和树莓派的支持，它主要采用交叉编译的方式，能够缩短开发周期，并在设备上快速运行。

4.2 TensorFlow 代码结构

让我们先下载 TensorFlow 的源代码。我们可以使用 Git 克隆项目：

```
$ git clone https://github.com/tensorflow/tensorflow.git
```

针对 TensorFlow 开发人员，可以使用 Git 查看历史记录并管理分支。还可以下载包含 TensorFlow 源代码的压缩文件（https://github.com/tensorflow/tensorflow/archive/master.zip）。

让我们来看一下 TensorFlow 的代码结构：

```
-rw-rw-r-- BUILD
drwxrwxr-x c
drwxrwxr-x cc
-rw-rw-r-- .clang-format
-rw-rw-r-- .blazerc
drwxrwxr-x compiler
drwxrwxr-x contrib
drwxrwxr-x core
drwxrwxr-x docs_src
drwxrwxr-x examples
drwxrwxr-x g3doc
drwxrwxr-x go
-rw-rw-r-- __init__.py
-rw-rw-r-- __init__.pyc
drwxrwxr-x java
drwxrwxr-x python
drwxrwxr-x stream_executor
-rw-rw-r-- tensorflow.bzl
-rw-rw-r-- tf_exported_symbols.lds
-rw-rw-r-- tf_version_script.lds
drwxrwxr-x tools
drwxrwxr-x user_ops
-rw-rw-r-- version_check.bzl
-rw-rw-r-- workspace.bzl
```

从上面代码可以看到，根目录中有两个重要的文件夹，一个是包含源代码的文件夹 core，另一个是 tools 文件夹。还有两个文件——.blazerc 和在 tools 文件夹下的 tf_env_collect.sh。其中.bazelrc 定义了 Bazel 配置，tf_env_collect.sh 用于收集系统信息，可以在提交 bug 时附加结果。

以下文件夹包含了 TensorFlow 实现，它支持 C、C++、Go、Java、Python。

```
drwxrwxr-x c
drwxrwxr-x cc
drwxrwxr-x compiler
drwxrwxr-x core
```

```
drwxrwxr-x go
drwxrwxr-x java
drwxrwxr-x python
```

docs_src 包括了文档文件：

```
drwxrwxr-x docs_src
```

examples 包括应用的例子，在这些例子中也包括 Android、iOS 的应用。

```
drwxrwxr-x examples
```

contrib 是一个特别的文件夹，它的作用在 README.md 里有说明：

Any code in this directory is not officially supported, and may change or be removed at any time without notice.

这段文字表明，此文件夹中的任何代码都不是官方的。如果想在产品中使用，就必须承担维护代码的责任，使其能够兼容。

在 contrib 中，有两个我们将频繁使用的文件夹。一个是 android，它是 TensorFlow Android 示例应用程序；另一个是 lite，它是 TensorFlow Lite。有趣的是，Java 不在 contrib 文件夹中。

Java 不在 contrib 文件夹中，是因为官方支持 TensorFlow 的 Java 接口。在 contrib/android 中的 TensorFlow Lite 和粘合代码不受官方支持。我们需要牢记这些。它现在可能不是很重要，但它可能会影响开发者未来的开发计划。

让我们来看一段 TensorFlow API 文档中的话：

TensorFlow has APIs available in several languages both for constructing and executing a TensorFlow graph. The Python API is at present the most complete and the easiest to use, but other language APIs may be easier to integrate into projects and may offer some performance advantages in graph execution.

A word of caution: the APIs in languages other than Python are not yet covered by the API stability promises.

由此可见，官方只支持 Python API。

我们先看看底层实现。TensorFlow 的底层都是利用 C 和 C++实现的。//tensorflow/cc:cc_ops 定义了自动生成的 C++接口，代码如下：

```
tf_gen_op_wrappers_cc(
    name = "cc_ops",
    api_def_srcs = ["//tensorflow/core/api_def:base_api_def"],
    op_lib_names = [
        "array_ops",
        "audio_ops",
        "candidate_sampling_ops",
        "control_flow_ops",
        "data_flow_ops",
        "image_ops",
        "io_ops",
        "linalg_ops",
        "logging_ops",
        "lookup_ops",
        "manip_ops",
        "math_ops",
        "nn_ops",
        "no_op",
        "parsing_ops",
        "random_ops",
        "sparse_ops",
        "state_ops",
        "string_ops",
        "training_ops",
        "user_ops",
    ],
    other_hdrs = [
        "ops/const_op.h",
        "ops/standard_ops.h",
    ],
    pkg = "//tensorflow/core",
)
```

TensorFlow C++参考中也清楚地记录了这些 Ops（Operations，算子）。TensorFlow 的 API 文档也是从代码中提取的。

让我们看一下 TensorFlow Android 的演示程序：

```
tf_cuda_library(
    name = "native",
    srcs = glob(["*.cc"]) + select({
        # The Android toolchain makes "jni.h" available in the include path.
        # For non-Android toolchains, generate jni.h and jni_md.h.
        "//tensorflow:android": [],
```

```
        "//conditions:default": [
            ":jni.h",
            ":jni_md.h",
        ],
    }),
    hdrs = glob(["*.h"]),
    copts = tf_copts() + [
        "-landroid",
        "-llog",
    ],
    includes = select({
        "//tensorflow:android": [],
        "//conditions:default": ["."],
    }),
    linkopts = [
        "-landroid",
    "-llog",
    ],
    deps = [
        **"//tensorflow/c:c_api",**
    ] + select({
        "//tensorflow:android": [
            **"//tensorflow/core:android_tensorflow_lib",**
        ],
        "//conditions:default": [
            "//tensorflow/core:all_kernels",
            "//tensorflow/core:direct_session",
            "//tensorflow/core:ops",
        ],
    }),
    alwayslink = 1,
)
//tensorflow/c:c_api 的定义
tf_cuda_library(
    name = "c_api",
    srcs = [
        "c_api.cc",
        "c_api_function.cc",
    ],
    hdrs = [
        "c_api.h",
    ],
    copts = tf_copts(),
    visibility = ["//visibility:public"],
```

```
    deps = select({
        "//tensorflow:android": [
            ":c_api_internal",
            "//tensorflow/core:android_tensorflow_lib_lite",
        ],
        "//conditions:default": [
        ],
    }) + select({
        "//tensorflow:with_xla_support": [
            "//tensorflow/compiler/tf2xla:xla_compiler",
            "//tensorflow/compiler/jit",
        ],
        "//conditions:default": [],
    }),
)
//tensorflow/core:android_tensorflow_lib 的定义
cc_library(
    name = "android_tensorflow_lib",
    srcs = if_android([":android_op_registrations_and_gradients"]),
    copts = tf_copts(),
    tags = [
        "manual",
        "notap",
    ],
    visibility = ["//visibility:public"],
    deps = [
        ":android_tensorflow_lib_lite",
        ":protos_all_cc_impl",
        "//tensorflow/core/kernels:android_tensorflow_kernels",
        "//third_party/eigen3",
        "@protobuf_archive//:protobuf",
    ],
    alwayslink = 1,
)
```

//tensorflow/core/kernels:android_tensorflow_kernels 基本包含了 Ops 的实现，它包括“//tensorflow/core/kernels:android_core_ops”和“//tensorflow/core/kernels:android_extended_ops”

```
cc_library(
    name = "android_tensorflow_kernels",
    srcs = select({
        "//tensorflow:android": [
            "//tensorflow/core/kernels:android_core_ops",
            "//tensorflow/core/kernels:android_extended_ops",
        ],
```

```
        "//conditions:default": [],
    }),
    copts = tf_copts(),
    linkopts = select({
        "//tensorflow:android": [
            "-ldl",
        ],
        "//conditions:default": [],
    }),
    tags = [
        "manual",
        "notap",
    ],
    visibility = ["//visibility:public"],
    deps = [
        "//tensorflow/core:android_tensorflow_lib_lite",
        "//tensorflow/core:protos_all_cc_impl",
        "//third_party/eigen3",
        "//third_party/fft2d:fft2d_headers",
        "@fft2d",
        "@gemmlowp",
        "@protobuf_archive//:protobuf",
    ],
    alwayslink = 1,
)
```

这个演示程序的构建目标是//tensorflow/examples/android:tensorflow_demo，这个构建目标不仅依赖于 Java JNI，而且也依赖于 Java 的原生实现（//tensorflow/java/src/main/native:native）。

4.3　构建及运行

在 1.10 版本之前，开发者需要手动编辑 WORKSPACE 设置工具链，代码如下：

```
android_sdk_repository(
    name = "androidsdk",
    api_level = 23,
    #确保你在 SDK 管理器中安装了 build_tools_version，因为它会定期更新
    build_tools_version = "26.0.1",
    # 替换为系统中 Android SDK 的路径
```

```
        **path = "/opt/Android/sdk",**
    )

    android_ndk_repository(
        name="androidndk",
        **path="/opt/Android/ndk",**
        # 编译 TensorFlow 需要将 api_level 设置为 14 或更高
        # 请注意，NDK 版本不是 API 级别
        api_level=14)
```

在新的版本里，这个手动的过程被改变了，解决办法是运行./configure，这个脚本会提示开发者设定环境变量：

```
    Please specify the location of python. [Default is /usr/bin/python]:

    Do you wish to build TensorFlow with XLA JIT support? [Y/n]:
    Do you wish to build TensorFlow with OpenCL SYCL support? [y/N]:
    Do you wish to build TensorFlow with ROCm support? [y/N]:
    Do you wish to build TensorFlow with CUDA support? [y/N]:
    Do you wish to download a fresh release of clang? (Experimental) [y/N]:
    Do you wish to build TensorFlow with MPI support? [y/N]:
    Please specify optimization flags to use during compilation when bazel option
"--config=opt" is specified [Default is -march=native -Wno-sign-compare]:
    Would you like to interactively configure ./WORKSPACE for Android builds?
[y/N]:
```

开发者只需要简单地回答问题。对于 Android，关键是设定工具链和 API Level，我们可以直接设置环境变量，configure 脚本用于读取这些变量，代码如下：

```
    export ANDROID_NDK_HOME=/opt/Android/ndk
    export ANDROID_SDK_HOME=/opt/Android/sdk
    export ANDROID_NDK_API_LEVEL=26
    export ANDROID_SDK_API_LEVEL=26
    export ANDROID_BUILD_TOOLS_VERSION=26.0.1
```

API Level 设成 26 即可支持 Android NNAPI。Android 的 SDK 中包含 NDK，可以直接用或下载单独的 NDK。另外，如果是公司的开发者，最好使用统一版本的 SDK 和 NDK。笔者见过一些公司，对此没有严格的限制，这其实存在潜在的危险。比如，NDK r14 产生的代码会有不充分优化的地方。笔者在完成从 r12 到 r16 的迁移时，将 SDK 和 NDK 安装在一个固定的路径。这样做有两个好处，一是可以实现远程自动管理，二是可以实现跨平台统一管理。

注：configure 不会安装 SDK 和 NDK，我们可以使用 Android Studio 中的 SDK Manager 来安装 SDK 和 NDK。

完成这些之后，我们就可以构建应用了，代码如下：

```
$ bazel build -c opt --config=android --cpu=arm64-v8a //tensorflow/examples/android:tensorflow_demo
$ adb install -r bazel-bin/tensorflow/examples/android/tensorflow_demo.apk
```

上面代码中的参数解释如下：

-c opt 让编译器优化代码。

--config=android 设置交叉编译器。最初，Bazel 仅在主机系统和目标系统相同的开发环境中进行编译。但是，对于嵌入式系统和移动系统，我们必须交叉编译代码。具体地说，开发者的工作站是 Intel/x86，但是我们想要生成的是可以运行 ARM 处理器的二进制代码。

--cpu=arm64-v8a 指定移动设备的 ABI。

你也可以使用更短的设定，下面来看看这个设定文件 tools/bazel.rc：

```
# Bazel 需要将--cpu 和--fat_apk_cpu 都设置为
# target CPU，以正确构建瞬态依赖项。读者可以参考
# https://docs.bazel.build/versions/master/user-manual.html
#flag--fat_apk_cpu
build:android --crosstool_top=//external:android/crosstool
build:android --host_crosstool_top=@bazel_tools//tools/cpp:toolchain
build:android --config=cross_compile
build:android_arm --config=android
build:android_arm --cpu=armeabi-v7a
build:android_arm --fat_apk_cpu=armeabi-v7a
build:android_arm64 --config=android
build:android_arm64 --cpu=arm64-v8a
build:android_arm64 --fat_apk_cpu=arm64-v8a
```

在笔者看来，开发者还可以使用 android_arm64 来实现相同的目标。如何找到用于开发手机的 abi 呢？

有两种方法，如果你在手机上启用了 USB 调试，则可以运行命令：

```
adb shell getprop | grep abi
```

你应该可以看到如下结果：

```
[ro.product.cpu.abilist]: [arm64-v8a,armeabi-v7a,armeabi]
[ro.product.cpu.abilist32]: [armeabi-v7a,armeabi]
[ro.product.cpu.abilist64]: [arm64-v8a]
```

最近的 Android 版本已停止支持 armeabi，对于不了解技术细节的用户，Android 构建并使用了所谓“胖”（Fat）APK 来解决这个问题。在“胖”APK 中，该软件包包含所有平台的所有可执行代码，这意味着在“胖”APK 中，它将包含 arm64-v8a、armeabi-v7a 甚至 Intel/x86 和 MIPS 的代码。

但是，它会使应用包更大。它不仅会导致存储问题，还需要更多空间来存储 APK，更大的包也意味着启动时间慢，消耗更多的电池电量。事实是，在 ARM 设备上，用户永远不会执行 Intel/x86 代码。因此，当我们从谷歌商店（Google Play Store）或其他 APK 商店下载 APK 时，这些服务将首先检测用户手机的 abi，然后将所谓的“瘦”APK 推送到我们的设备。“瘦”APK 仅包含一个 abi 的代码。所以请记住，如果开发者为一个 abi 编写优化代码，则必须为其他 abi 准备类似的优化。

接下来，安装 APK：

```
$ adb install bazel-bin/tensorflow/examples/android/tensorflow_demo.apk
```

现在，在连接到桌面的手机上，我们会看到一个图标，单击该图标可以运行它。如果有多台设备连接到主机，则需要按序列号选择设备。例如，列出连接到主机的设备：

```
$ adb devices
List of devices attached
adb server is out of date.  killing...
* daemon started successfully *
 UYT0217C05001756device
```

通过下面的命令安装演示程序：

```
$ adb install -s UYT0217C05001756device bazel-bin/tensorflow/examples/
android/tensorflow_demo.apk
```

如果你喜欢使用 Android Studio 中的即时运行功能，Bazel 也有类似的功能，称为移动安装。我们不必运行 Bazel 构建和安装，可以执行下面的命令：

```
$ adb mobile-install -s UYT0217C05001756device bazel-bin/tensorflow/
examples/android/tensorflow_demo.apk
```

mobile-install 从以下三个方面来改善构建和安装：

- Sharded dexing。在构建应用程序的 Java 代码后，Bazel 将类文件分为大小大致相等的若干片，并在各个分片上单独调用 dex。在上次构建后未更改的分片可以不调用 dex。

- 增量文件传输。Android 资源，.dex 文件和本机库将从主.apk 中删除，并存储在单独的移动安装目录下。这样可以独立更新代码和 Android 资源，而不需要重新安装整个应用程序。因此，传输文件所花费的时间更少，只有已更改的.dex 文件才会在设备上重新编译。
- 从 APK 外部加载应用程序的部分内容。将一个小的存根（stub）应用程序放入 APK 中，从设备上的移动安装目录加载 Android 资源、Java 代码和本机代码，然后将控制权转移给实际的应用程序。除下面描述的一些极端情况外，这对应用程序来说都是透明的。

建议开发者使用这个功能。在安装完成后，你应该会在手机上看到四个应用图标，如图 4-1 所示。

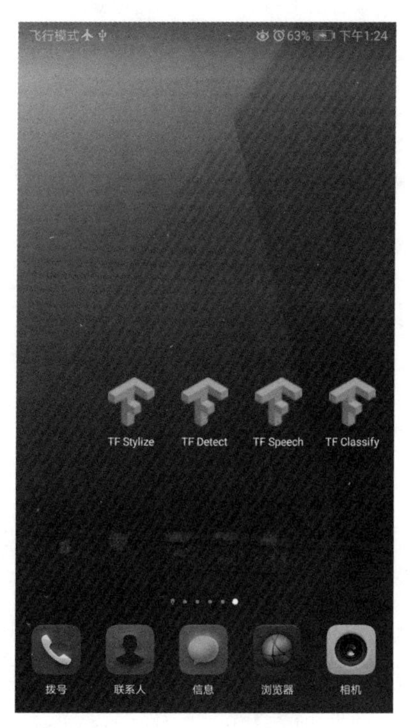

图 4-1　应用安装

现在让我们看一下 examples/android/AndroidManifest.xml 这个文件：

```
<activity android:name="org.tensorflow.demo.ClassifierActivity"
```

```
        android:screenOrientation="portrait"
        android:label="@string/activity_name_classification">
    <intent-filter>
        <action android:name="android.intent.action.MAIN" />
        <category android:name="android.intent.category.LAUNCHER" />
        <category android:name="android.intent.category.LEANBACK_LAUNCHER" />
    </intent-filter>
</activity>

<activity android:name="org.tensorflow.demo.DetectorActivity"
        android:screenOrientation="portrait"
        android:label="@string/activity_name_detection">
    <intent-filter>
        <action android:name="android.intent.action.MAIN" />
        <category android:name="android.intent.category.LAUNCHER" />
        <category android:name="android.intent.category.LEANBACK_LAUNCHER" />
    </intent-filter>
</activity>

<activity android:name="org.tensorflow.demo.StylizeActivity"
        android:screenOrientation="portrait"
        android:label="@string/activity_name_stylize">
    <intent-filter>
        <action android:name="android.intent.action.MAIN" />
        <category android:name="android.intent.category.LAUNCHER" />
        <category android:name="android.intent.category.LEANBACK_LAUNCHER" />
    </intent-filter>
</activity>

<activity android:name="org.tensorflow.demo.SpeechActivity"
    android:screenOrientation="portrait"
    android:label="@string/activity_name_speech">
    <intent-filter>
        <action android:name="android.intent.action.MAIN" />
        <category android:name="android.intent.category.LAUNCHER" />
        <category android:name="android.intent.category.LEANBACK_LAUNCHER" />
    </intent-filter>
</activity>
```

从上面代码中可以了解到，一次安装命令安装了四个应用程序，TF Stylize、TF Detect、TF Speech、TF Classify。我们可以手动单击图标来启动应用，也可以运行类似的操作来启动应用：

```
    $ adb shell am start -n org.tensorflow.demo/org.tensorflow.demo.
ClassifierActivity
    $ adb shell am start -n org.tensorflow.demo/.ClassifierActivity
```

如果你不能在办公室工作，那么你可以通过在主机上使用键盘来简化与手机的互动。当你在家工作时，它非常方便。人们喜欢的另一种方式是使用 adb port forward，你可以在本地编译 APK 并安装在远程手机上。笔者认为第一种方式非常强大而且速度很快，当你在家工作时，甚至不需要触摸手机，是不是很酷？

4.3.1　代码的流程

现在，读者应该能够构建和运行这些 TensorFlow 应用程序了。现在让我们来解释和理解代码，以及学习如何在 Android 应用程序中使用 TensorFlow。

如表 4-1 所示是 Java 的执行步骤。

表 4-1　Java 的执行步骤

代码逻辑	代码路径
应用逻辑	tensorflow/examples/android
TensorFlow 外包层	tensorflow/contrib/android
TensorFlow java 层	tensorflow/java

TensorFlow 提供了许多示例来帮助用户了解 TensorFlow API 的用法。读者可以从 label_image 示例开始。它有三个版本，一个是用 Python 编写的，一个是用 C++编写的，一个是用 Java 编写的。

推理的基本代码流程如图 4-2 所示。

在 Python 中，读者可能会发现“import /”是输入和输出层的前缀：

```
input_name = "import/" + input_layer
output_name = "import/" + output_layer
```

这是因为 import_graph_def 加上了前缀：

```
def load_graph(model_file):
  graph = tf.Graph()
  graph_def = tf.GraphDef()

  with open(model_file, "rb") as f:
    graph_def.ParseFromString(f.read())
  with graph.as_default():
    **tf.import_graph_def(grap_def)**

  return graph
```

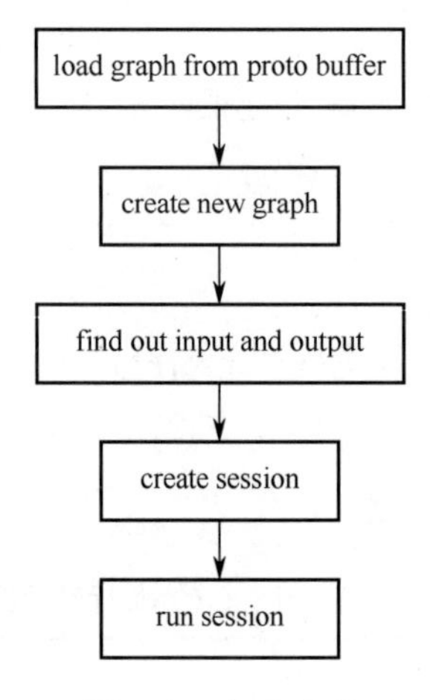

图 4-2　代码流程

我们可以把上面高亮的行写成：

```
**tf.import_graph_def(graph_def, name="")**
```

这种修改可能更适合匹配 C++和 Java 示例中的逻辑，因此 Python、C++和 Java 看起来非常相似。

4.3.2　代码的依赖性

理解代码的依赖性是理解代码的重要一环，我们可以通过执行下面的命令来列出针对 tensorflow_demo 这个应用的所有依赖关系，并把依赖关系输出到 graph.dot 文件中：

```
$ bazel query --noimplicit_deps 'deps(//tensorflow/examples/android:tensorflow_demo)' --output graph > graph.dot
```

然后，使用 xdot 打开文件：

```
$ xdot graph.dot
```

在 Mac 上，先安装 xdot，命令如下：

```
brew install xdot
```

上面的命令将构建一个相当大的依赖图，这个图里包含了构建目标的所有依赖关系，图形文件会很大，读者可以自己生成看一下。我们可以从这个图体会到 Blaze 强调的对每一个依赖的确定性。读者还可以将 graphviz 的 dot 文件转换为 jpg 或 png 图像文件，但是，在 xdot 中加载或转换成图像文件需要很长时间。有一些技术可以进一步优化 graphviz 输出文件，但它不是本书的主题。

4.3.3　性能和代码跟踪

在演示代码中，Android Trace 已被 Trace 函数启用，下面的代码展示了如何使用 Trace：

```
// 将输入数据复制到 TensorFlow
Trace.beginSection("feed");
inferenceInterface.feed(inputName, floatValues, 1, inputSize, inputSize, 3);
Trace.endSection();

// 运行推理调用
Trace.beginSection("run");
inferenceInterface.run(outputNames, logStats);
Trace.endSection();

// 将输出张量复制回输出数组
Trace.beginSection("fetch");
inferenceInterface.fetch(outputName, outputs);
Trace.endSection();
```

让我们运行上面的代码并检查其性能：

```
$ python /opt/Android/sdk/platform-tools/systrace/systrace.py --app org.tensorflow.demo -t 5 -o result.html
```

在 Android Trace 运行中，我们需要指定应用程序名称为 org.tensorflow.demo，收集跟踪数据 5 秒，输出文件为 result.html。

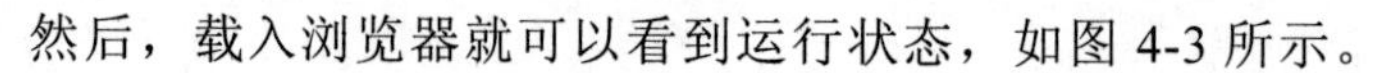

然后，载入浏览器就可以看到运行状态，如图 4-3 所示。

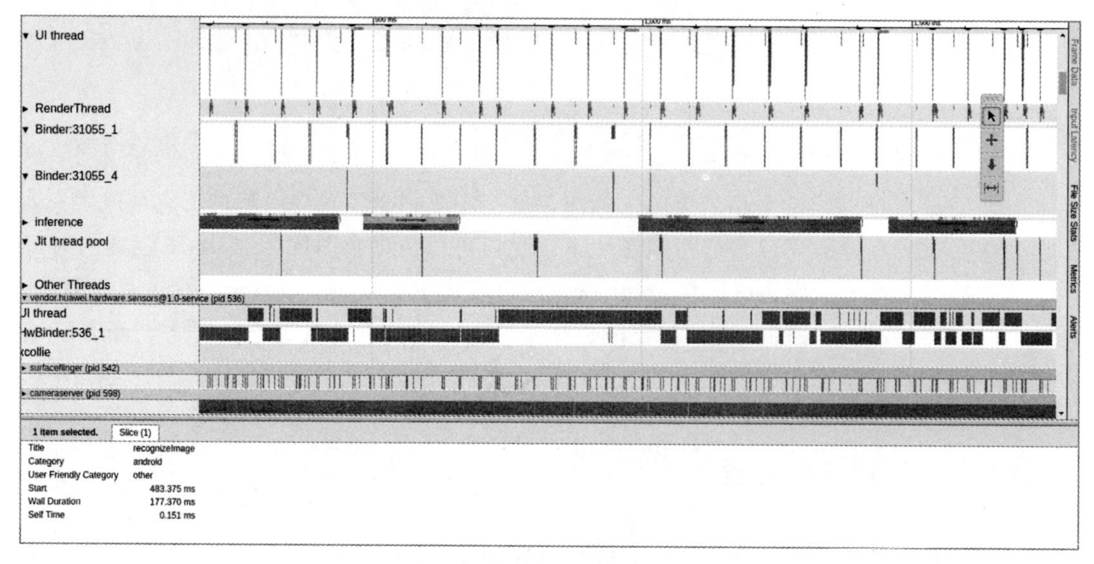

图 4-3 运行状态

第5章 用TensorFlow Mobile构建机器学习应用

本章将介绍怎样使用 TensorFlow 开发 Android 的机器学习应用。

5.1 准备工作

为了看到更多日志记录，可以在 tensorflow/examples/android/src/org/tensorflow/demo/env/Logger.java 中把 Log.DEBUG 换成 Log.VERBOSE：

```
private static final String DEFAULT_TAG = "tensorflow";
private static final int DEFAULT_MIN_LOG_LEVEL = Log.DEBUG;
```

在应用中要用到模型和标注表，通常我们希望开发者能自动下载编译，应用也能自动读取，我们来看一下 Bazel 是怎样实现这个功能的。首先我们看一下 tensorflow/

examples/android/下的 build 文件：

```
android_binary( name = "tensorflow_demo",
assets = [
    "//tensorflow/examples/android/assets:asset_files",
    ":external_assets",
],
```

build 文件里定义了编译目标 tensorflow_demo，以及编译要生成的应用 tensorflow_demo.apk。这个应用依赖于 asset 中的:extemal assets，它在 build 文件中的定义如下：

```
filegroup(
    name = "external_assets",
    srcs = [
        "@inception_v1//:model_files",
        "@mobile_ssd//:model_files",
        "@speech_commands//:model_files",
        "@stylize//:model_files",
    ],
)
```

filegroup 定义了一组文件，这些文件就是模型和标注文件。tensorflow_demo 这个应用会把其中定义的四个目标包含的文件全部保存到 APK 里的 assert 下面。由于模型文件占用存储空间较大，所以通常只保存所需模型即可。现在看看其中的一个构建目标“@inception_v1//:model_files”，它的定义可以在 workspace 文件里找到，workspace 文件代码如下：

```
http_archive(
    name = "inception_v1",
    build_file = "//:models.BUILD",
    sha256    =    "7efe12a8363f09bc24d7b7a450304a15655a57a7751929b2c1593a
71183bb105",
    urls = [
        "http://storage.googleapis.com/download.tensorflow.org/models
/inception_v1.zip",
        "http://download.tensorflow.org/models/inception_v1.zip",
    ],
)
```

Bazel 会自动从 http://storage.googleapis.com/download.tensorflow.org/models/inception_v1.zip 或 http://download.tensorflow.org/models/inception_v1.zip 下载解压文件，并检查 sha256 值。下载成功后，使用"//:models.BUILD"进行构建，具体实现代码如下：

```
filegroup(
    name = "model_files",
    srcs = glob(
        [
            "**/*",
        ],
        exclude = [
            "**/BUILD",
            "**/WORKSPACE",
            "**/LICENSE",
            "**/*.zip",
        ],
    ),
)
```

这段代码会自动忽略下载文件中的一些文件，比如 BUILD、WORKSPACE 等，而把其他文件作为构建目标，这样其他构建目标可以参照这些文件。

比如，我们可以通过下面的命令下载模型文件：

```
$ wget "http://storage.googleapis.com/download.tensorflow.org/models/inception_v1.zip"
$ unzip inception_v1.zip
$ ls -all
-r--r----- 1    10492 Nov 18  2015 imagenet_comp_graph_label_strings.txt
-rw-r----- 1 49937249 Jan 22  2018 inception_v1.zip
-r--r----- 1    11416 Nov 18  2015 LICENSE
-rw-r----- 1 53881635 Sep 28  2017 tensorflow_inception_graph.pb
```

得到这些文件并成功构建 tensorflow_demo.apk 后，我们可以执行下面的命令：

```
$ unzip bazel-bin/tensorflow/examples/android/tensorflow_demo.apk
$ ls -all asserts
-rw-rw-rw- 1      328 Jan  1  2010 BUILD.bazel
-rw-rw-rw- 1      665 Jan  1  2010 coco_labels_list.txt
-rw-rw-rw- 1  3771239 Jan  1  2010 conv_actions_frozen.pb
-rw-rw-rw- 1       60 Jan  1  2010 conv_actions_labels.txt
-rw-rw-rw- 1    10492 Jan  1  2010 imagenet_comp_graph_label_strings.txt
-rw-rw-rw- 1    11416 Jan  1  2010 LICENSE
-rw-rw-rw- 1 29083865 Jan  1  2010 ssd_mobilenet_v1_android_export.pb
-rw-rw-rw- 1   563897 Jan  1  2010 stylize_quantized.pb
-rw-rw-rw- 1 53881635 Jan  1  2010 tensorflow_inception_graph.pb
drwxr-x--- 2     4096 Jan 21 11:01 thumbnails
```

```
-rw-rw-rw- 1        28 Jan  1  2010 WORKSPACE
```

我们定义的模型文件和标注目标文件都被保存在 asserts 下面了。请注意，tensorflow_inception_graph.pb 文件大小近 54MB，stylize_quantized.pb 文件大小不到 600KB。模型占用的存储空间还是不小的。APK 的大小，不仅影响了对设备存储的要求，而且，用户第一次下载要花费大量时间，这对用户体验也有很大影响。

5.2 图像分类（Image Classification）

图像分类是人工智能的一个主要应用，我们来看一下怎样在移动设备上实现图像的分类。

5.2.1 应用

下面的例子主要讲解 tensorflow/examples/android/src/org/tensorflow/demo 下的 TensorFlowImageClassifier.java 文件。

图像分类的 Activity 是 tensorflow/examples/android/src/org/tensorflow/demo/ ClassifierActivity.java。它的定义是：

```
public class ClassifierActivity extends CameraActivity implements OnImageAvailableListener {}
```

它继承了 CameraActivity，并实现了 OnImageAvailableListener。CameraActivity 实现了 Android 应用的基本生命周期的功能，比如 onStart、onCreate、onStop、onDestroy 等。

另外，它实现了相机的预览（Preview）。实现预览的主要原因是，我们要从相机里取得图像的数据。在 onCreate 里首先调用 setFragment，在这段代码里会生成 CameraConnectionFragment 的一个实例。

```
if (useCamera2API) {
  CameraConnectionFragment camera2Fragment =
      CameraConnectionFragment.newInstance(
          new CameraConnectionFragment.ConnectionCallback() {
            @Override
            public void onPreviewSizeChosen(final Size size, final int
rotation) {
              previewHeight = size.getHeight();
```

```
                previewWidth = size.getWidth();
                CameraActivity.this.onPreviewSizeChosen(size, rotation);
              }
            },
            this,
            getLayoutId(),
            getDesiredPreviewFrameSize());

      camera2Fragment.setCamera(cameraId);
      fragment = camera2Fragment;
    } else {
      fragment =
          new LegacyCameraConnectionFragment(this, getLayoutId(),
getDesiredPreviewFrameSize());
    }
```

tensorflow/examples/android/src/org/tensorflow/demo 下的 CameraConnectionFragment. Java 文件实现了 CameraConnectionFragment，这是 Android 的一个 Fragment。其关键的功能是由 setUpCameraOutputs 实现的，代码如下：

```
      private void setUpCameraOutputs() {
        final Activity activity = getActivity();
        final   CameraManager   manager   =   (CameraManager)   activity.
getSystemService(Context.CAMERA_SERVICE);
        try {
          final    CameraCharacteristics    characteristics    =    manager.
getCameraCharacteristics(cameraId);

          final StreamConfigurationMap map =
              characteristics.get(CameraCharacteristics.SCALER_STREAM_
CONFIGURATION_MAP);

          // 使用最大尺寸进行图像抓取
          final Size largest =
              Collections.max(
                  Arrays.asList(map.getOutputSizes(ImageFormat.YUV_ 420_888)),
                  new CompareSizesByArea());

          sensorOrientation   =   characteristics.get(CameraCharacteristics.
SENSOR_ORIENTATION);

          // 预览尺寸过大会超过相框
```

```
        // 垃圾捕获数据
        previewSize =
            chooseOptimalSize(map.getOutputSizes(SurfaceTexture.class),
                inputSize.getWidth(),
                inputSize.getHeight());

        // TextureView 的长宽比要与我们选择的预览尺寸相匹配
        final int orientation = getResources().getConfiguration().
orientation;
        if (orientation == Configuration.ORIENTATION_LANDSCAPE) {
          textureView.setAspectRatio(previewSize.getWidth(),
previewSize.getHeight());
        } else {
          textureView.setAspectRatio(previewSize.getHeight(),
previewSize.getWidth());
        }
      } catch (final CameraAccessException e) {
        LOGGER.e(e, "Exception!");
      } catch (final NullPointerException e) {
        // 当此代码运行的设备不支持 Camera2API 时，抛弃 NPE
        // 此代码运行的设备

        ErrorDialog.newInstance(getString(R.string.camera_error))
            .show(getChildFragmentManager(), FRAGMENT_DIALOG);
        throw new RuntimeException(getString(R.string.camera_error));
      }

      cameraConnectionCallback.onPreviewSizeChosen(previewSize,
sensorOrientation);
    }
```

在通过 getCameraCharacteristics(cameraId)获得相机的属性后，通过与图像的解像度进行对比，调用 chooseOptimalSize 找到最合适的预览 Preview 的尺寸。

```
  protected static Size chooseOptimalSize(final Size[] choices, final int
width, final int height) {
    final int minSize = Math.max(Math.min(width, height),
MINIMUM_PREVIEW_SIZE);
    final Size desiredSize = new Size(width, height);

    // 收集支持的分辨率，这些分辨率至少与预览图面一样大
    boolean exactSizeFound = false;
```

```
      final List<Size> bigEnough = new ArrayList<Size>();
      final List<Size> tooSmall = new ArrayList<Size>();
      for (final Size option : choices) {
        if (option.equals(desiredSize)) {
          // 设置尺寸，但不返回，以便记录剩余尺寸
          exactSizeFound = true;
        }

        if (option.getHeight() >= minSize && option.getWidth() >= minSize) {
          bigEnough.add(option);
        } else {
          tooSmall.add(option);
        }
      }

      LOGGER.i("Desired size: " + desiredSize + ", min size: " + minSize + "x"
+ minSize);
      LOGGER.i("Valid preview sizes: [" + TextUtils.join(", ", bigEnough) + "]");
      LOGGER.i("Rejected preview sizes: [" + TextUtils.join(", ", tooSmall) +
"]");

      if (exactSizeFound) {
        LOGGER.i("Exact size match found.");
        return desiredSize;
      }

      // 选择其中最小的一个图像
      if (bigEnough.size() > 0) {
        final    Size    chosenSize    =    Collections.min(bigEnough,    new
CompareSizesByArea());
        LOGGER.i("Chosen   size:   "   +   chosenSize.getWidth()   +   "x"   +
chosenSize.getHeight());
        return chosenSize;
      } else {
        LOGGER.e("Couldn't find any suitable preview size");
        return choices[0];
      }
    }
```

如果找到完全匹配的图像就返回，否则返回较小的图像。

到这里我们基本上就实现了把一个相机设定好并取得其预览的功能。由于我们在这里

需要构造一个完整的应用，所以要通过大量代码来实现从一个实际的应用中获取图像程序的功能。在测试或者非应用的程序中可以选用一些静态的图像，代码也会简单很多。

回到 CameraActivity.java 的 setFragment，下面的代码会显示 Fragment：

```
getFragmentManager()
     .beginTransaction()
     .replace(R.id.container, fragment)
     .commit();
```

在 onResume()里面，主要做了两件事，一是建立一个后台线程，二是启动相机。启动相机就是调用 Android CameraManager 的 openCamera 函数，这个应用做了一个简单的封装，代码如下：

```
@Override
  public void onResume() {
    super.onResume();
    startBackgroundThread();

    // 当屏幕关闭并重新打开时，surfacetexture 为可用状态，并且不会调用
"onsurfaceextureavailable"，我们可以打开一个摄像头进行预览
    if (textureView.isAvailable()) {
      openCamera(textureView.getWidth(), textureView.getHeight());
    } else {
      textureView.setSurfaceTextureListener(surfaceTextureListener);
    }
  }
```

建立后台线程。过程很简单，这里调用 Android 的 Handler 和 HandlerThread，生成并启动一个名为“ImageListener”的线程。

```
private void startBackgroundThread() {
   backgroundThread = new HandlerThread("ImageListener");
   backgroundThread.start();
   backgroundHandler = new Handler(backgroundThread.getLooper());
 }
```

启动相机。如果相机被打开和启动，那么下面的函数就会被调用，并启动 createCamera-PreviewSession。

```
private final CameraDevice.StateCallback stateCallback =
    new CameraDevice.StateCallback() {
```

```
        @Override
        public void onOpened(final CameraDevice cd) {
          // 当相机打开时，这个方法就会被调用，我们即可开始预览相机

          cameraOpenCloseLock.release();
          cameraDevice = cd;
          createCameraPreviewSession();
        }
```

使用相机预览 Preview 的应用，基本要实现两个功能，一是设定相机的预览大小，二是实现一个相机录入会话 CameraCaptureSession，它的实现在

private void createCameraPreviewSession()中。具体实现过程如下：

首先，建立一个肌理（Texture）和与其关联的 TextureView，作为图像输出的显示区，这样我们就可以在设备上看到预览图像：

```
    final SurfaceTexture texture = textureView.getSurfaceTexture();
    assert texture != null;

    // 我们将默认缓冲区的大小配置为所需的相机预览大小
    texture.setDefaultBufferSize(previewSize.getWidth(),
previewSize.getHeight());

    // 这是我们需要开始预览的输出曲面
    final Surface surface = new Surface(texture);

    // 我们用输出曲面设置了 CaptureRequest.Builder
    previewRequestBuilder                                              =
cameraDevice.createCaptureRequest(CameraDevice.TEMPLATE_PREVIEW);
    previewRequestBuilder.addTarget(surface);
```

然后，新建一个 ImageReader，并设定预览大小和图像数据的回调：

```
    // 为预览帧创建读卡器
    previewReader =
        ImageReader.newInstance(
            previewSize.getWidth(), previewSize.getHeight(), ImageFormat.
YUV_420_888, 2);

    previewReader.setOnImageAvailableListener(imageListener,
```

```
backgroundHandler);
    previewRequestBuilder.addTarget(previewReader.getSurface());
```

最后，调用 cameraDevice.createCaptureSession 生成一个相机的图像抓取会话，关键的地方是生成一个会话的回调，代码如下：

```
      public        void        onConfigured(final        CameraCaptureSession
cameraCaptureSession) {
          // 相机已关闭
          if (null == cameraDevice) {
            return;
          }
           // 当会话准备就绪时，开始显示预览
           captureSession = cameraCaptureSession;
          try {
            // 在相机预览时，自动对焦是连续的
            previewRequestBuilder.set(
                CaptureRequest.CONTROL_AF_MODE,
                CaptureRequest.CONTROL_AF_MODE_CONTINUOUS_PICTURE);
            // 必要时自动启用闪存
            previewRequestBuilder.set(
                CaptureRequest.CONTROL_AE_MODE, CaptureRequest.CONTROL_AE_
MODE_ON_AUTO_FLASH);
              // 显示相机预览
            previewRequest = previewRequestBuilder.build();
            captureSession.setRepeatingRequest(
                previewRequest, captureCallback, backgroundHandler);
          } catch (final CameraAccessException e) {
            LOGGER.e(e, "Exception!");
          }
        }
```

至此，这个应用的基本准备功能和框架都有了，可以从相机看到图像，图像数据通过回调函数也可以得到。下面把应用和图像分类的模型联系起来。

ClassifierActivity 实现了 OnImageAvailableListener，也实现了 Camera.PreviewCallback，当新的图像被相机生成后，onImageAvailable 会被调用。这个函数主要做两个工作，一是图像格式的转换，二是调用 processImage()进行图像处理。

```
    @Override
```

```
public void onImageAvailable(final ImageReader reader)
```

下面是函数的定义。它的图像转换主要调用 ImageUtils 类里的函数。

```
Trace.beginSection("imageAvailable");
final Plane[] planes = image.getPlanes();
fillBytes(planes, yuvBytes);
yRowStride = planes[0].getRowStride();
final int uvRowStride = planes[1].getRowStride();
final int uvPixelStride = planes[1].getPixelStride();

imageConverter =
    new Runnable() {
      @Override
      public void run() {
        ImageUtils.convertYUV420ToARGB8888(
            yuvBytes[0],
            yuvBytes[1],
            yuvBytes[2],
            previewWidth,
            previewHeight,
            yRowStride,
            uvRowStride,
            uvPixelStride,
            rgbBytes);
      }
    };
```

然后，调用 processImage()函数：

```
Trace.beginSection("imageAvailable");
Trace.endSection();
```

该函数使用 Android 的 Trace 功能，可以记录图像处理所需要的时间。

请注意，在 ImageUtils 里有几个 YUV 和 RGB 转换的函数。YUV 在硬件图像处理里使用较多，多数相机的输出格式也支持 YUV。RGB 是一种历史很长的格式，它代表了红绿蓝在颜色里的构成，比较容易理解，很多应用会使用这种格式。所以，不同颜色的转换是必要的，比如 convertYUV420ToARGB8888，相关转换代码如下：

```
int yp = 0;
for (int j = 0; j < height; j++) {
```

```
    int pY = yRowStride * j;
    int pUV = uvRowStride * (j >> 1);

    for (int i = 0; i < width; i++) {
      int uv_offset = pUV + (i >> 1) * uvPixelStride;

      out[yp++] = YUV2RGB(
          0xff & yData[pY + i],
          0xff & uData[uv_offset],
          0xff & vData[uv_offset]);
    }
  }
```

对于每一个像素，依据下面的公式进行转换：

```
private static int YUV2RGB(int y, int u, int v) {
  // nR = (int)(1.164 * nY + 2.018 * nU);
  // nG = (int)(1.164 * nY - 0.813 * nV - 0.391 * nU);
  // nB = (int)(1.164 * nY + 1.596 * nV);
}
```

细心的读者一定会注意到，这个函数很简单，是简单的 for 循环和加乘法的组合，也一定会消耗很多时间，如果是 512×512 的图像，会重复很多简单计算。在图像处理中，通过优化函数来提高性能的方法有几种，可以用机器原生语言 C 或者汇编来实现，也可以使用机器上的硬件加速来实现。

实际上，在实测中我们也发现了这个问题，但是这是一个演示应用，并不过多占有机器性能，因此就采用了现在这个方案。

这个应用会得到两个回调，一是相机本身产生的图像，二是预览 Preview 的图像，我们会根据应用采用不同的处理方式。

为了实现图像分类，在 ClassifierActivity 里重载了 processImage，并调用分类模型，输出分类的结果：

```
    @Override
    protected void processImage() {
      rgbFrameBitmap.setPixels(getRgbBytes(),   0,   previewWidth,   0,   0,
previewWidth, previewHeight);
      final Canvas canvas = new Canvas(croppedBitmap);
      canvas.drawBitmap(rgbFrameBitmap, frameToCropTransform, null);
```

```
        // 检查当前的 TF 输入
        if (SAVE_PREVIEW_BITMAP) {
          ImageUtils.saveBitmap(croppedBitmap);
        }
        runInBackground(
            new Runnable() {
              @Override
              public void run() {
                final long startTime = SystemClock.uptimeMillis();
                final   List<Classifier.Recognition>   results   =   classifier.
recognizeImage(croppedBitmap);
                lastProcessingTimeMs = SystemClock.uptimeMillis() - startTime;
                LOGGER.i("Detect: %s", results);
                cropCopyBitmap = Bitmap.createBitmap(croppedBitmap);
                if (resultsView == null) {
                  resultsView = (ResultsView) findViewById(R.id.results);
                }
                resultsView.setResults(results);
                requestRender();
                readyForNextImage();
              }
            });
      }
```

classifier.recognizeImage 实现了图像分类，它的输入是一个 BitMap，然后返回一个分类结果的队列：

```
    final  List<Classifier.Recognition>  results  =  classifier.recognizeImage
(croppedBitmap);
```

由于模型的预测需要耗费计算资源和时间，这个函数一定要运行在非主线程上，这里使用了 Runnable。分类的结果是一个类 Recognition，包含了分类的结果，它的定义如下：

```
    public class Recognition {
      // 已识别内容的唯一标识符。特定于类，而不是对象
     private final String id;
      // 显示识别名称
     private final String title;
      // 一个可排序的分数，表示相对于其他人的认可度，该认可度值越高越好
     private final Float confidence;
```

```
    // 源图像中可用于识别对象位置的可选位置
    private RectF location;
```

title 是被分类物体的名称，confidence 是返回的与分类物体相对应的信心数，这个数值越高越好。

下面是图像分类的实现过程，它由几部分构成。

第一步，先把图像的 RGB 数值转换为浮点值：

```
    bitmap.getPixels(intValues, 0, bitmap.getWidth(), 0, 0, bitmap.getWidth(),
bitmap.getHeight());
    for (int i = 0; i < intValues.length; ++i) {
      final int val = intValues[i];
      floatValues[i * 3 + 0] = (((val >> 16) & 0xFF) - imageMean) / imageStd;
      floatValues[i * 3 + 1] = (((val >> 8) & 0xFF) - imageMean) / imageStd;
      floatValues[i * 3 + 2] = ((val & 0xFF) - imageMean) / imageStd;
    }
```

第二步，把浮点值传入模型中进行预测：

```
    inferenceInterface.feed(inputName, floatValues, 1, inputSize, inputSize, 3);
```

第三步，运行模型的会话：

```
    inferenceInterface.run(outputNames, logStats);
```

第四步，取出模型的预测结果：

```
    inferenceInterface.fetch(outputName, outputs);
```

以上几步，除了第一步，基本和用 Python 实现预测的步骤是一样的，只不过是用 Java 实现的。

下一步是对输出的结果进行比较，最后输出的处理比较简单，一是比较信心值，只显示几个比较有意义的结果，二是省去其他信心值比较低的结果。具体实现代码如下：

```
    // 找到最佳分类
    PriorityQueue<Recognition> pq =
        new PriorityQueue<Recognition>(
            3,
            new Comparator<Recognition>() {
              @Override
```

```
          public int compare(Recognition lhs, Recognition rhs) {
            // 故意颠倒，以在队列的最前面建立高度的信心
    return Float.compare(rhs.getConfidence(), lhs.getConfidence());
          }
        });
    for (int i = 0; i < outputs.length; ++i) {
      if (outputs[i] > THRESHOLD) {
        pq.add(
            new Recognition(
              "" + i, labels.size() > i ? labels.get(i) : "unknown",
outputs[i], null));
      }
    }
```

在程序中对每一个步骤都使用了 Trace，我们可以使用工具更好地了解每一步消耗的时间，并根据结果做优化。

```
    Trace.beginSection("feed");
    Trace.endSection();
```

5.2.2　模型

这个应用的模型是 tensorflow_inception_graph.pb，它由下面的代码定义。

```
    private static final String MODEL_FILE = "file:///android_asset/
tensorflow_inception_graph.pb";
      private static final String LABEL_FILE =
        "file:///android_asset/imagenet_comp_graph_label_strings.txt";
```

模型的源文件代码如下：

```
    http_archive(
        name = "inception_v1",
        build_file = "//:models.BUILD",
        sha256 = "7efe12a8363f09bc24d7b7a450304a15655a57a7751929b2c1593
a71183bb105",
        urls = [
            "http://storage.googleapis.com/download.tensorflow.org/
models/inception_v1.zip",
            "http://download.tensorflow.org/models/inception_v1.zip",
```

```
    ],
)
```

我们可以下载并解压这个模型，得到 tensorflow_inception_graph.pb，然后运行如下代码：

```
    $ python tensorflow/python/tools/import_pb_to_tensorboard.py --model_dir
tensorflow_inception_graph.pb --log_dir /tmp/log
    $ tensorboard --logdir /tmp/log
```

打开浏览器，输入地址 http://localhost:6006，可以看到 Inception 模型图，如图 5-1 所示。

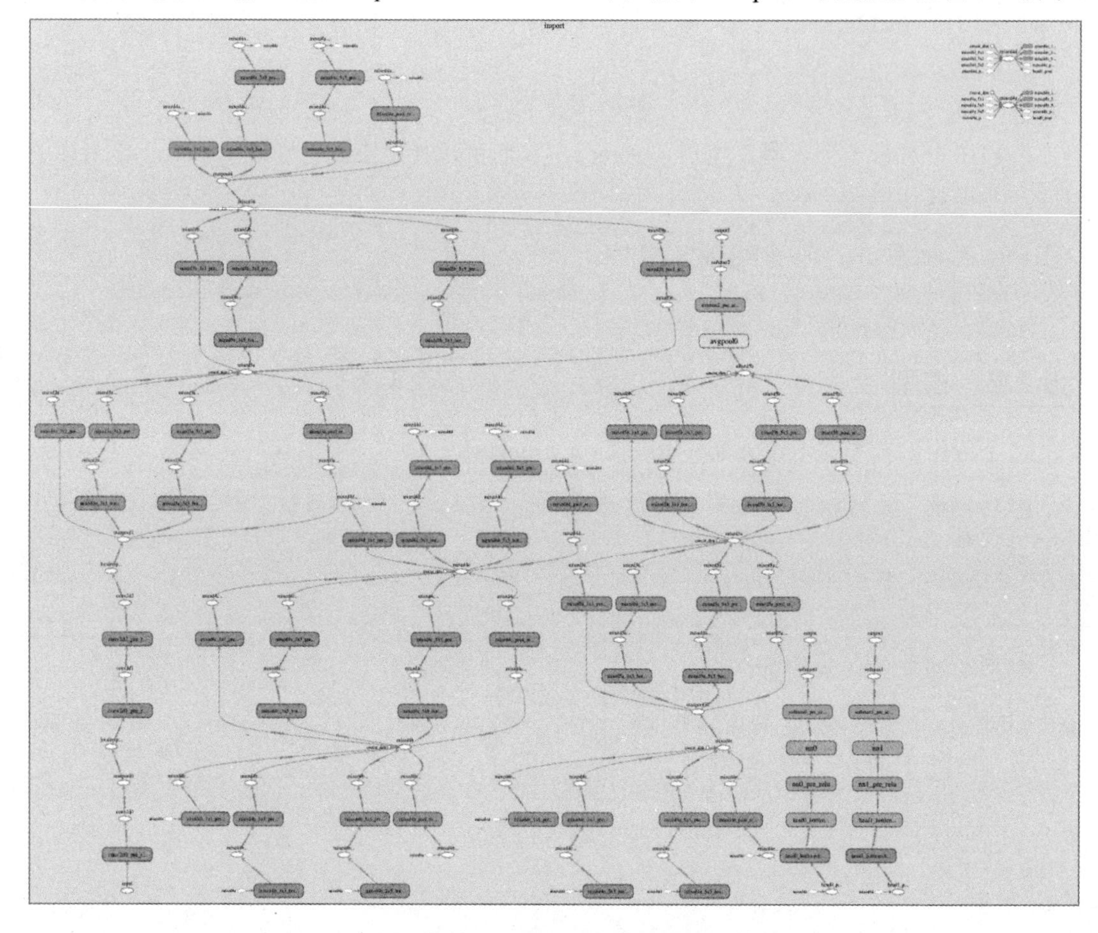

图 5-1　Inception 模型图

5.3 物体检测（Object Detection）

5.3.1 应用

DetectorActivity.java 实现了使用机器学习模型来进行物体检测。这个应用只是对相机的图像进行处理，所以只继承了 OnImageAvailableListener：

```
    public class DetectorActivity extends CameraActivity implements
OnImageAvailableListener
```

这个例子提供了三种模式，它们的定义如下：

```
    private enum DetectorMode {
      TF_OD_API, MULTIBOX, YOLO;
    }
```

默认值是 YOLO，读者可以改变这个值，重新编译运行这个应用，比较三个模型的差异。运行这三个模型的方式是基本一致的。

但是，对于信心（confidence）值，不同的模型会有不同处理，应用里分别定义了相应的数值：

```
    // 跟踪检测的最小置信度
    private static final float MINIMUM_CONFIDENCE_TF_OD_API = 0.6f;
    private static final float MINIMUM_CONFIDENCE_MULTIBOX = 0.1f;
    private static final float MINIMUM_CONFIDENCE_YOLO = 0.25f;
```

另外，模型对于输入图像有不同的要求，它们输入的变量名也不同。而且，应用把模型用不同的文件名保存起来，下面是各自的定义。这些常量其实可以封装起来，由 MODE 来决定，代码看起来会更简单一些。

```
    private static final int MB_INPUT_SIZE = 224;
    private static final int MB_IMAGE_MEAN = 128;
    private static final float MB_IMAGE_STD = 128;
    private static final String MB_INPUT_NAME = "ResizeBilinear";
    private static final String MB_OUTPUT_LOCATIONS_NAME = "output_
locations/Reshape";
    private static final String MB_OUTPUT_SCORES_NAME = "output_
scores/Reshape";
    private static final String MB_MODEL_FILE =
```

```
"file:///android_asset/multibox_model.pb";
    private static final String MB_LOCATION_FILE =
        "file:///android_asset/multibox_location_priors.txt";

    private static final int TF_OD_API_INPUT_SIZE = 300;
    private static final String TF_OD_API_MODEL_FILE =
        "file:///android_asset/ssd_mobilenet_v1_android_export.pb";
    private static final String TF_OD_API_LABELS_FILE = "file:
///android_asset/coco_labels_list.txt";

    private static final String YOLO_MODEL_FILE = "file:///android_
asset/graph-tiny-yolo-voc.pb";
    private static final int YOLO_INPUT_SIZE = 416;
    private static final String YOLO_INPUT_NAME = "input";
    private static final String YOLO_OUTPUT_NAMES = "output";
    private static final int YOLO_BLOCK_SIZE = 32;
```

以下代码生成了 MultiBoxTracker。MultiBoxTracker 负责追踪检测物体并把物体的外框（Box）表示出来。DetectorActivity 会把物体检测的结果传进来，并由 MultiBoxTracker 显示到设备上。

```
    tracker = new MultiBoxTracker(this);
```

下面的代码根据 MODE 生成了检测器的实例。这三个检测器都实现了接口 public interface Classifier，所以都被封装了起来，使用者也不用关心它们的实现细节。

```
    if (MODE == DetectorMode.YOLO) {
      detector =
         TensorFlowYoloDetector.create(
            getAssets(),
            YOLO_MODEL_FILE,
            YOLO_INPUT_SIZE,
            YOLO_INPUT_NAME,
            YOLO_OUTPUT_NAMES,
            YOLO_BLOCK_SIZE);
      cropSize = YOLO_INPUT_SIZE;
     } else if (MODE == DetectorMode.MULTIBOX) {
      detector =
         TensorFlowMultiBoxDetector.create(
            getAssets(),
            MB_MODEL_FILE,
            MB_LOCATION_FILE,
            MB_IMAGE_MEAN,
            MB_IMAGE_STD,
```

```
            MB_INPUT_NAME,
            MB_OUTPUT_LOCATIONS_NAME,
            MB_OUTPUT_SCORES_NAME);
      cropSize = MB_INPUT_SIZE;
    } else {
      try {
        detector = TensorFlowObjectDetectionAPIModel.create(
            getAssets(), TF_OD_API_MODEL_FILE, TF_OD_API_LABELS_FILE, TF_OD_
API_INPUT_SIZE);
        cropSize = TF_OD_API_INPUT_SIZE;
    }
```

在图像识别（Image Classification）的例子中，我们关心被识别的物体是什么，在这个应用里，我们也关心物体的位置，所以在接口 Classifier 里还定义了一个类型为 android.graphics.RectF 的位置变量：

```
// 源图像中可用于识别对象位置的可选位置
private RectF location;
```

Tracker 和 detector 完成了这个应用的主要功能。Detector 负责调用模型，返回检测到的物体名称和位置。Tracker 负责把检测到的物体显示到显示器上。在这个应用里，为了显示和追踪物体会花费大量的代码，包括 MultiBoxTracker.java、ObjectTracker.java 和位于 tensorflow/examples/android/jni/object_tracking 下的大量 C++代码。

模型会返回检测到的物体名称和位置。但是，返回的物体名称和位置不是连续的，如果我们只是按照模型返回的数值进行显示是绝对不够的，在现实中一般物体的移动速度是有限的，移动的位置是有关联性的。按照这个原理，我们可以通过代码实现简单的物体追踪。在机器学习的应用里，实际上编写大量的代码还是为了能实现应用的逻辑功能。

这个应用的主要功能是在 processImage()里实现的，分别调用了 Tracker 和 Detector 的函数。一个简化的实现过程大概有三个步骤：

```
@Override
 protected void processImage() {
   // 第一步，清除已检测的物体，准备下一次显示
   ++timestamp;
   tracker.onFrame(
      previewWidth,
      previewHeight,
      getLuminanceStride(),
      sensorOrientation,
      originalLuminance,
      timestamp);
```

```
      trackingOverlay.postInvalidate();

      runInBackground(
          new Runnable() {
            @Override
            public void run() {
              // 第二步，调用模型，获取被识别物体的队列，包括物体的名称和位置
              LOGGER.i("Running detection on image " + currTimestamp);
              final long startTime = SystemClock.uptimeMillis();
              final        List<Classifier.Recognition>        results        =
detector.recognizeImage(croppedBitmap);
              lastProcessingTimeMs = SystemClock.uptimeMillis() - startTime;

              // 第三步，把已识别的物体显示出来
              tracker.trackResults(mappedRecognitions,          luminanceCopy,
currTimestamp);
              trackingOverlay.postInvalidate();

              requestRender();
              computingDetection = false;
            }
          });
    }
```

我们先看一下 detector 的实现。上面提到了，这个应用支持三个检测器，三个检测器可实现接口 classifier，它们的使用方法比较类似，都实现了 recognizeImage，但也有些不同。以下是三个检测器检测的具体实现方式，读者可以看看它们的不同。

TensorFlowMultiBoxDetector.java

```
    Trace.beginSection("preprocessBitmap");
    // 对图像数据进行预处理，转化为标准化浮点数
    bitmap.getPixels(intValues, 0, bitmap.getWidth(), 0, 0, bitmap.getWidth(),
bitmap.getHeight());

    for (int i = 0; i < intValues.length; ++i) {
      floatValues[i * 3 + 0] = (((intValues[i] >> 16) & 0xFF) - imageMean) /
imageStd;
      floatValues[i * 3 + 1] = (((intValues[i] >> 8) & 0xFF) - imageMean) /
imageStd;
      floatValues[i * 3 + 2] = ((intValues[i] & 0xFF) - imageMean) / imageStd;
    }
    Trace.endSection();
    // 位图预处理
```

```
    // 将输入数据复制到 TensorFlow 中
    Trace.beginSection("feed");
    inferenceInterface.feed(inputName, floatValues, 1, inputSize, inputSize, 3);
    Trace.endSection();

    // 运行推理调用
    Trace.beginSection("run");
    inferenceInterface.run(outputNames, logStats);
    Trace.endSection();

    // 将输出张量复制回输出数组
    Trace.beginSection("fetch");
    final float[] outputScoresEncoding = new float[numLocations];
    final float[] outputLocationsEncoding = new float[numLocations * 4];
    inferenceInterface.fetch(outputNames[0], outputLocationsEncoding);
    inferenceInterface.fetch(outputNames[1], outputScoresEncoding);
    Trace.endSection();
```

TensorFlowObjectDetectionAPIModel.java：

```
    Trace.beginSection("preprocessBitmap");
    // 预处理图像数据，从 0x00rrggbb 格式的 int 中提取 r、g 和 b 字节
    bitmap.getPixels(intValues, 0, bitmap.getWidth(), 0, 0, bitmap.getWidth(),
bitmap.getHeight());

    for (int i = 0; i < intValues.length; ++i) {
      byteValues[i * 3 + 2] = (byte) (intValues[i] & 0xFF);
      byteValues[i * 3 + 1] = (byte) ((intValues[i] >> 8) & 0xFF);
      byteValues[i * 3 + 0] = (byte) ((intValues[i] >> 16) & 0xFF);
    }
    Trace.endSection();
    // 位图预处理
    // 将输入数据复制到 TensorFlow 中
    Trace.beginSection("feed");
    inferenceInterface.feed(inputName, byteValues, 1, inputSize, inputSize, 3);
    Trace.endSection();

    // 运行推理调用
    Trace.beginSection("run");
    inferenceInterface.run(outputNames, logStats);
    Trace.endSection();
```

TensorFlowYoloDetector.java：

```
    Trace.beginSection("preprocessBitmap");
    // 对图像数据进行预处理，转化为标准化浮点数
```

```
    bitmap.getPixels(intValues, 0, bitmap.getWidth(), 0, 0, bitmap.getWidth(),
bitmap.getHeight());

    for (int i = 0; i < intValues.length; ++i) {
      floatValues[i * 3 + 0] = ((intValues[i] >> 16) & 0xFF) / 255.0f;
      floatValues[i * 3 + 1] = ((intValues[i] >> 8) & 0xFF) / 255.0f;
      floatValues[i * 3 + 2] = (intValues[i] & 0xFF) / 255.0f;
    }
    Trace.endSection(); //位图预处理
    // 将输入数据复制到 TensorFlow 中
    Trace.beginSection("feed");
    inferenceInterface.feed(inputName, floatValues, 1, inputSize, inputSize, 3);
    Trace.endSection();

    timer.endSplit("ready for inference");

    // 运行推理调用
    Trace.beginSection("run");
    inferenceInterface.run(outputNames, logStats);
    Trace.endSection();
```

在 TensorFlowObjectDetectionAPIModel 里，模型的数据输入类型是 byte，而 TensorFlowYoloDetector 和 TensorFlowMultiBoxDetector 的输入类型是 float，一个是定点数，另一个是浮点数。

5.3.2 模型

这个应用用到了三个模型，其中 inception 模型在上面已经介绍了。现在来看看另外两个，一个是 Mobile Net。

```
    http_archive(
        name = "mobile_ssd",
        build_file = "//:models.BUILD",
        sha256 = "bddd81ea5c80a97adfac1c9f770e6f55cbafd7cce4d3bbe15fbeb041e6b8
f3e8",
        urls = [
            "http://storage.googleapis.com/download.tensorflow.org/models/
object_detection/ssd_mobilenet_v1_android_export.zip",
            "http://download.tensorflow.org/models/object_detection/ssd_
mobilenet_v1_android_export.zip",
        ],
    )
```

使用前面介绍过的过程，模型结构的视图如图 5-2 所示。

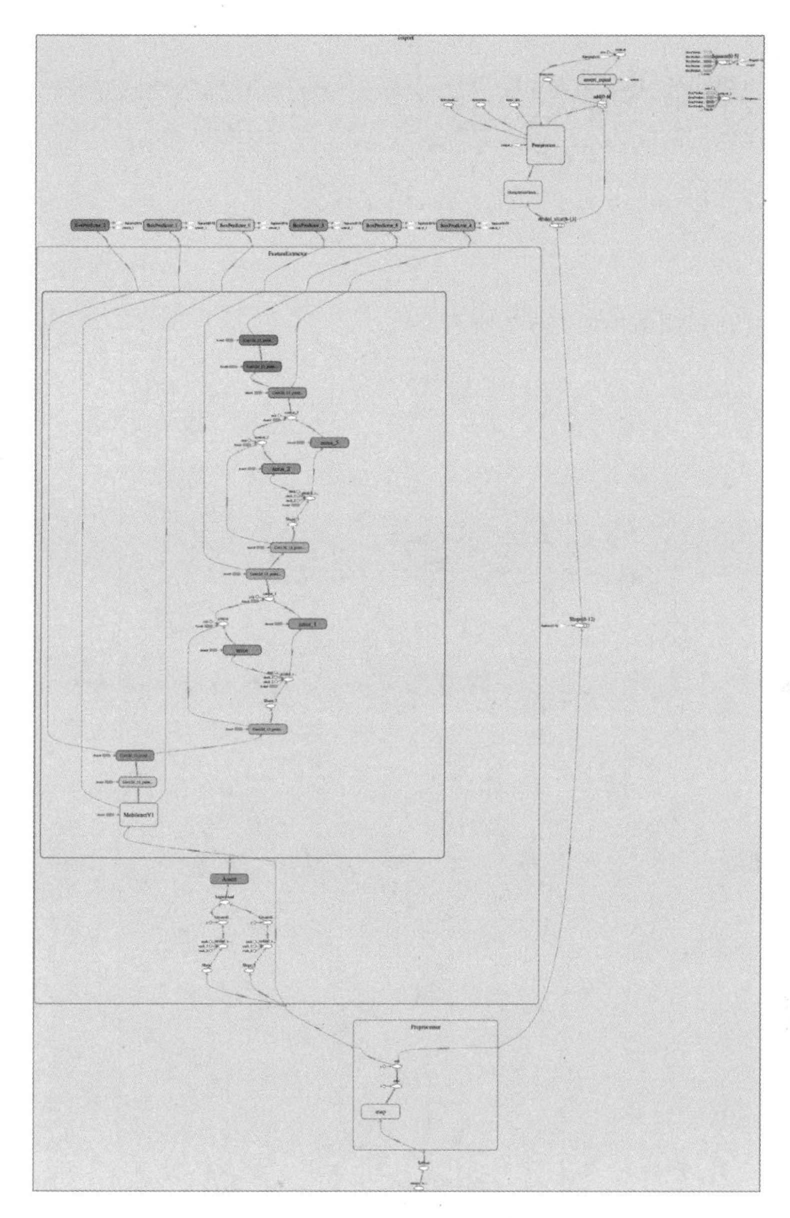

图 5-2　SSD Mobile Net 模型图

另外一个是 Mobile Multibox 模型，代码如下：

```
http_archive(
    name = "mobile_multibox",
    build_file = "//:models.BUILD",
```

```
        sha256                                                                  =
"859edcddf84dddb974c36c36cfc1f74555148e9c9213dedacf1d6b613ad52b96",
        urls = [
            "http://storage.googleapis.com/download.tensorflow.org/models/
mobile_multibox_v1a.zip",
            "http://download.tensorflow.org/models/mobile_multibox_v1a.zip",
        ],
    )
```

模型的可视化视图结构如图 5-3 所示。

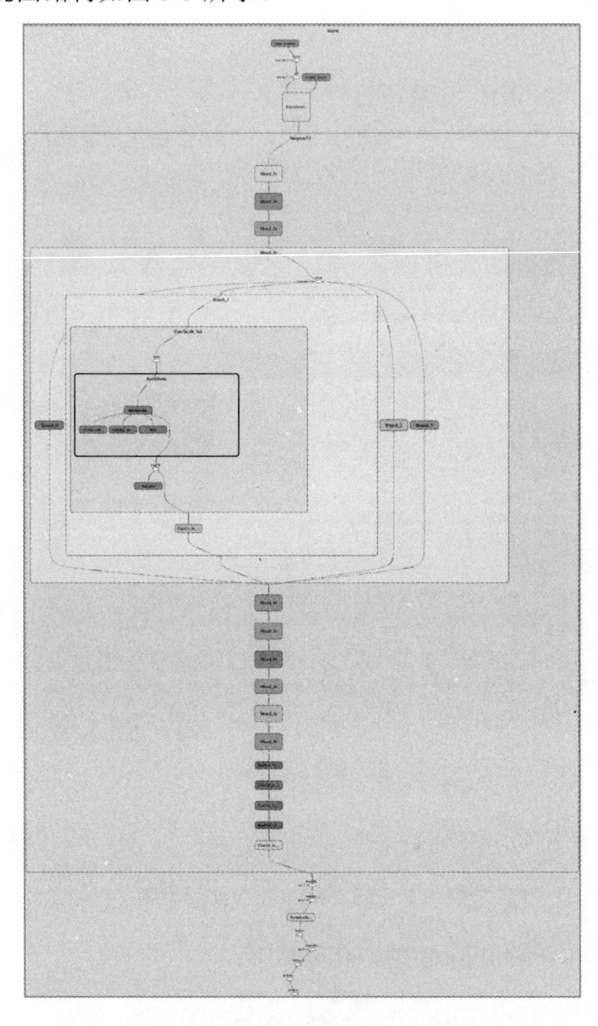

图 5-3　Mobile Multibox 结构图

5.4　时尚渲染（Stylization）

5.4.1　应用

时尚渲染的应用是由 StylizeActivity.java 实现的。主要的功能也是由 processImage()调用 stylizeImage()来实现的。

```
    ++frameNum;
    bitmap.getPixels(intValues, 0, bitmap.getWidth(), 0, 0, bitmap.getWidth(),
bitmap.getHeight());

    for (int i = 0; i < intValues.length; ++i) {
      final int val = intValues[i];
      floatValues[i * 3] = ((val >> 16) & 0xFF) / 255.0f;
      floatValues[i * 3 + 1] = ((val >> 8) & 0xFF) / 255.0f;
      floatValues[i * 3 + 2] = (val & 0xFF) / 255.0f;
    }

    // 将输入数据复制到 TensorFlow
    LOGGER.i("Width: %s , Height: %s", bitmap.getWidth(), bitmap.getHeight());
    inferenceInterface.feed(
        INPUT_NODE, floatValues, 1, bitmap.getWidth(), bitmap.getHeight(), 3);
    inferenceInterface.feed(STYLE_NODE, styleVals, NUM_STYLES);

    inferenceInterface.run(new String[] {OUTPUT_NODE}, isDebug());
    inferenceInterface.fetch(OUTPUT_NODE, floatValues);

    for (int i = 0; i < intValues.length; ++i) {
      intValues[i] =
          0xFF000000
              | (((int) (floatValues[i * 3] * 255)) << 16)
              | (((int) (floatValues[i * 3 + 1] * 255)) << 8)
              | ((int) (floatValues[i * 3 + 2] * 255));
    }

    bitmap.setPixels(intValues, 0, bitmap.getWidth(), 0, 0, bitmap.getWidth(),
bitmap.getHeight());
```

实现的过程也相对简单，通过调用 inferenceInterface 的 feed、run、fetch 后得到像素的浮点值，然后转换成对应的 RGB 的定点数值即可。

5.4.2 模型

时尚模型源文件的定义如下：

```
    http_archive(
        name = "stylize",
        build_file = "//:models.BUILD",
        sha256  =  "3d374a730aef330424a356a8d4f04d8a54277c425e274ecb7d9c83aa
912c6bfa",
        urls = [
            "http://storage.googleapis.com/download.tensorflow.org/models/
stylize_v1.zip",
            "http://download.tensorflow.org/models/stylize_v1.zip",
        ],
    )
```

我们把模型文件下载下来，按照上面的例子，转换成一个可视化的图形，代码如下：

```
    $ python tensorflow/python/tools/import_pb_to_tensorboard.py --model_dir
stylize_quantized.pb --log_dir /tmp/log
```

被细化的 Style 模型如图 5-4 所示。

5.5 声音识别（Speech Recognization）

5.5.1 应用

以上的几个例子都是和图像处理有关的，本节这个例子是和声音有关的，也是典型的机器学习的例子。这个应用会识别声音，并把命令显示出来。由于要使用麦克风，所以要把权限加上：

```
    <uses-permission android:name="android.permission.RECORD_AUDIO" />
```

这个应用由 tensorflow/examples/android/src/org/tensorflow/demo/SpeechActivity.java 实现。当应用启动的时候，同时启动麦克风的录音和声音识别，代码的实现方法如下：

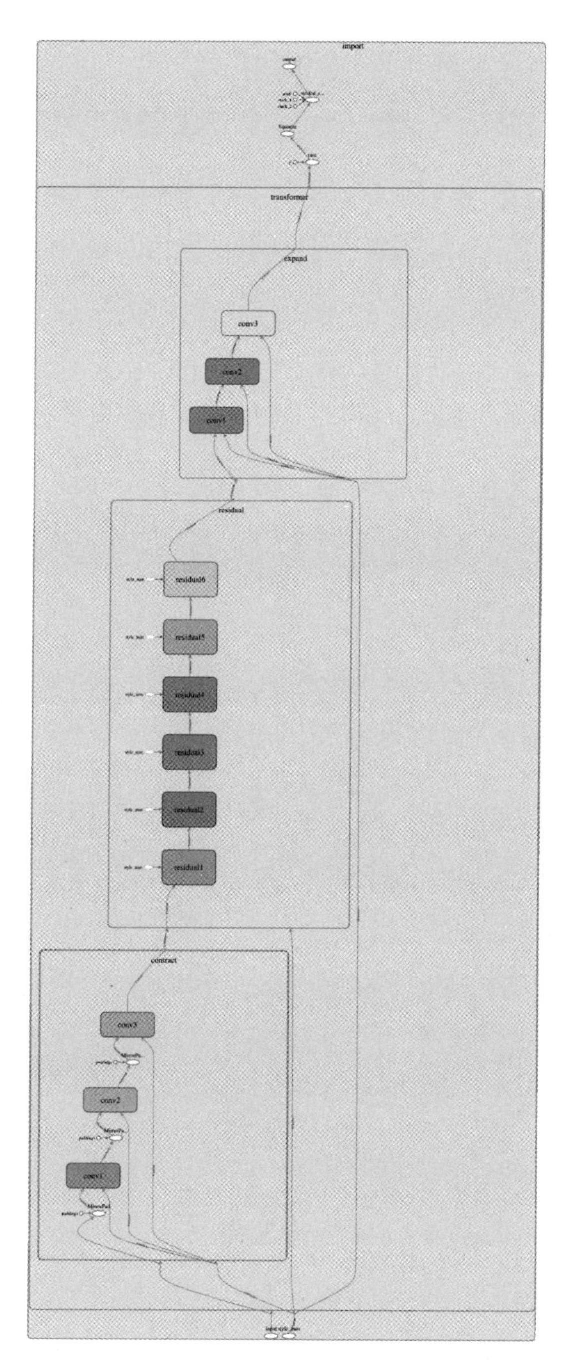

图 5-4　被细化的 Style 模型图

```
    protected void onCreate(Bundle savedInstanceState) {
      // 加载 TensorFlow 模型
      inferenceInterface   =   new   TensorFlowInferenceInterface(getAssets(),
MODEL_FILENAME);

      // 启动录制并识别线程
      requestMicrophonePermission();
      startRecording();
      startRecognition();
    }
```

startRecording 实现的主要功能是，启动一个线程。在这个线程里，首先设定录音设备，然后启动录音。在录音开始后，把数据存到一个缓存数组里，供声音识别使用。代码的简单实现方法如下：

```
      private void record() {
        android.os.Process.setThreadPriority(android.os.Process.THREAD_
PRIORITY_AUDIO);

        // 预估设备需要的缓冲区大小
        int bufferSize =
            AudioRecord.getMinBufferSize(
                SAMPLE_RATE, AudioFormat.CHANNEL_IN_MONO, AudioFormat.ENCODING_
PCM_16BIT);
        if  (bufferSize  ==  AudioRecord.ERROR  ||  bufferSize  ==  AudioRecord.
ERROR_BAD_VALUE) {
          bufferSize = SAMPLE_RATE * 2;
        }
        short[] audioBuffer = new short[bufferSize / 2];

        AudioRecord record =
            new AudioRecord(
                MediaRecorder.AudioSource.DEFAULT,
                SAMPLE_RATE,
                AudioFormat.CHANNEL_IN_MONO,
                AudioFormat.ENCODING_PCM_16BIT,
                bufferSize);

        record.startRecording();

        // 循环并收集音频数据并将其复制到循环缓冲区
          while (shouldContinue) {
          int numberRead = record.read(audioBuffer, 0, audioBuffer.length);
          int maxLength = recordingBuffer.length;
          int newRecordingOffset = recordingOffset + numberRead;
          int secondCopyLength = Math.max(0, newRecordingOffset - maxLength);
          int firstCopyLength = numberRead - secondCopyLength;
```

```
        //存储所有数据，以便识别线程访问
        //线程将从这个缓冲区被复制到自己的缓冲区中，这个过程是加锁的
        recordingBufferLock.lock();
        try {
          System.arraycopy(audioBuffer, 0, recordingBuffer, recordingOffset,
firstCopyLength);
          System.arraycopy(audioBuffer, firstCopyLength, recordingBuffer, 0,
secondCopyLength);
          recordingOffset = newRecordingOffset % maxLength;
        } finally {
          recordingBufferLock.unlock();
        }
      }
    }
```

注意，代码设定的声音格式是 SAMPLE_RATE = 16000，声音取样的频率是 16k，采用的一个声道是 CHANNEL_IN_MONO，声音样本的数据格式是 16bit 和 AudioFormat.ENCODING_PCM_16BIT。声音识别主要是由 startRecognition()和 recognize()实现的，也是在另外一个线程上实现的。实现的主要代码如下：

```
  // 输入介于-1.0f 和 1.0f 之间的浮点值
    for (int i = 0; i < RECORDING_LENGTH; ++i) {
      floatInputBuffer[i] = inputBuffer[i] / 32767.0f;
    }

    // 运行模型
    inferenceInterface.feed(SAMPLE_RATE_NAME, sampleRateList);
    inferenceInterface.feed(INPUT_DATA_NAME, floatInputBuffer, RECORDING_
LENGTH, 1);
    inferenceInterface.run(outputScoresNames);
    inferenceInterface.fetch(OUTPUT_SCORES_NAME, outputScores);
```

首先，要做数据类型的转换，把 16 位的定点数转换为浮点数。然后，还是在调用 feed、run 和 fetch 后，把预测的结果转换为命令显示出来。

5.5.2　模型

时尚模型源文件的定义如下：

```
  http_archive(
      name = "speech_commands",
      build_file = "//:models.BUILD",
      sha256 = "c3ec4fea3158eb111f1d932336351edfe8bd515bb6e87aad4f25dbad0a
600d0c",
```

```
        urls = [
            "http://storage.googleapis.com/download.tensorflow.org/models/
speech_commands_v0.01.zip",
            "http://download.tensorflow.org/models/speech_commands_
v0.01.zip",
        ],
    )
```

我们把模型文件下载下来，按照上面的例子，转换成一个可视化的图形，代码如下：

```
    $ python tensorflow/python/tools/import_pb_to_tensorboard.py --model_dir
conv_actions_frozen.pb --log_dir /tmp/log
```

模型的视图如图 5-5 所示。

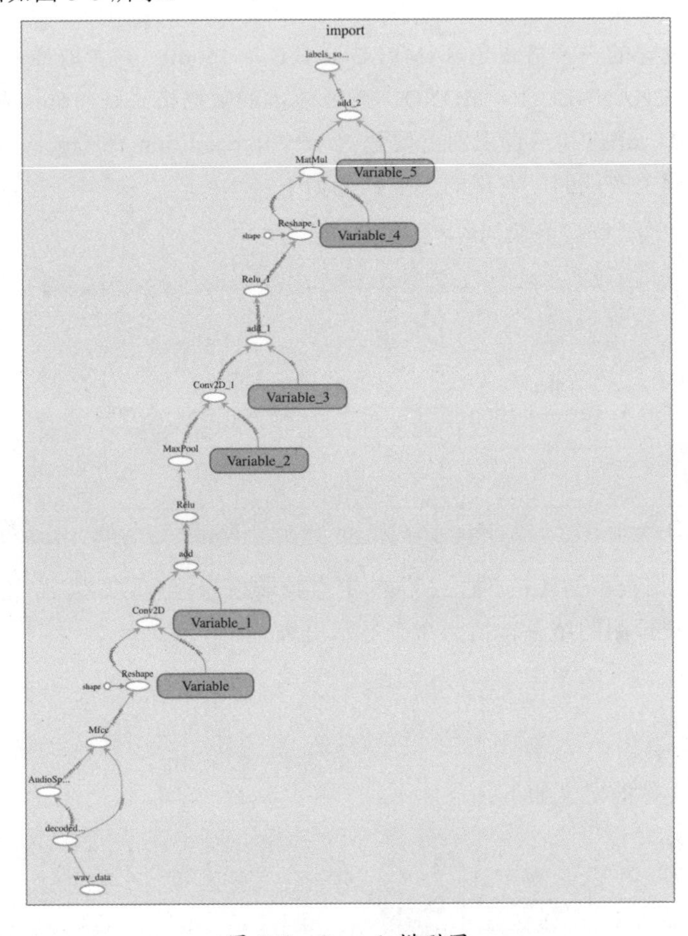

图 5-5　Speech 模型图

第 6 章 TensorFlow Lite 的架构

本章将介绍 TensorFlow Lite 的架构。TensorFlow Lite 的架构如图 6-1 所示。

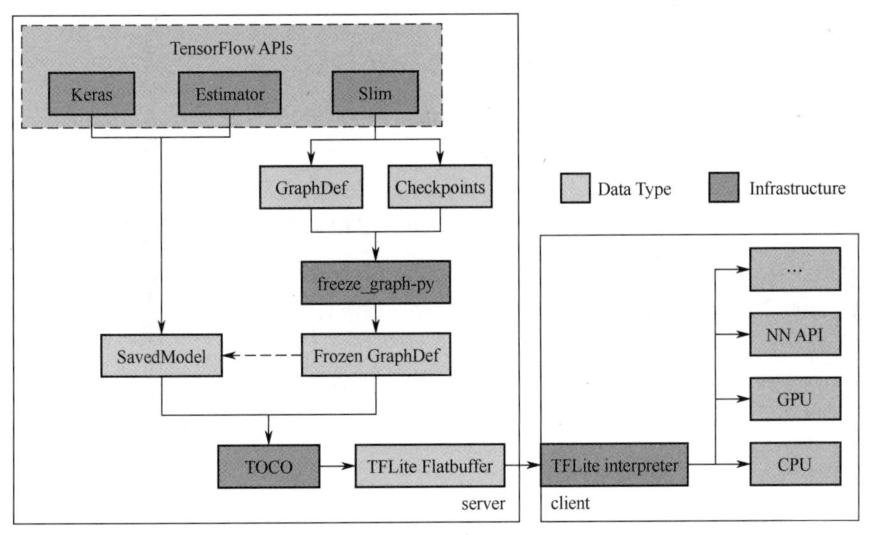

图 6-1　TensorFlow Lite 的架构

下面介绍 TensorFlow 的主要技术特点和开发者要了解的基本知识。

6.1 模型格式

首先需要理解 TensorFlow 的数据格式和它们的技术特点。

6.1.1 Protocol Buffers

Google Protocol Buffers(简称 Protobuf)是谷歌公司内部的混合语言数据标准,Protobuf 是一种轻便、高效的结构化数据存储格式,可以用于结构化数据串行化或序列化。它很适合做数据存储或 RPC 数据交换格式,可用于通信协议、数据存储等领域的与语言无关、平台无关、可扩展的序列化结构数据格式。它提供了包括 C++、Java、Python 在内的多种语言的 API。

Protobuf 几乎可以称为谷歌数据格式的基础,是每一个谷歌工程师必须掌握的技能,而且它也在不断地进化当中。有意思的是,谷歌的团队还在不断改善它的性能,可以想象在数据中心里,如果常用到这种数据结构,那么它的性能被提高即使是千分之一、万分之一,对整个系统性能的提高也是非常明显的。

Protobuf 有 Version 2 和 Version 3。然而,不能说 Version 3 一定比 Version 2 好,现在有很多代码还在使用 Version 2。但是,在 TensorFlow 里用到的都是 Version 3:

```
syntax = "proto3";
```

我们看一下 TensorFlow 的几个 Protobuf,来了解一下它的定义,也了解一下 TensorFlow 里的一些基本概念。

```
// 描述机器学习模型训练或推理输入数据示例的协议消息

syntax = "proto3";

import "third_party/tensorflow/core/example/feature.proto";
option cc_enable_arenas = true;
option java_outer_classname = "ExampleProtos";
option java_multiple_files = true;
option java_package = "org.tensorflow.example";
```

```
// 使用 copybara 在外部添加 go 包
package tensorflow;

message Example {
  Features features = 1;
};
```

这段代码定义了 Protobuf 本身的包名称 TensorFlow 和它对应 Java 包的名称 org.tensorflow.example。Protobuf 定义了 Example，只包含一个成员 features，它的定义在文件 third_party/tensorflow/core/example/feature.proto 中，代码如下：

```
syntax = "proto3";
option cc_enable_arenas = true;
option java_outer_classname = "FeatureProtos";
option java_multiple_files = true;
option java_package = "org.tensorflow.example";
// 使用 copybara 在外部添加 go 包
package tensorflow;

// 保存重复基本值的容器
message BytesList {
  repeated bytes value = 1;
}
message FloatList {
  repeated float value = 1 [packed = true];
}
message Int64List {
  repeated int64 value = 1 [packed = true];
}

// 非顺序数据容器
message Feature {
  // 每个功能都可以是一种
  oneof kind {
    BytesList bytes_list = 1;
    FloatList float_list = 2;
    Int64List int64_list = 3;
  }
};

message Features {
  // 从要素名称映射到要素
```

```
  map<string, Feature> feature = 1;
};
```

Features 就是字符串到 Feature 的映射，而每个 Feature 只能是下面三个中的一个，ByteList、FloatList 或 Int64List。

另外，我们可以看到在 Protobuf 里有大量的注释，这是一个好的工作习惯，同时也能帮助我们了解这些定义的用法，很多文档就是从这些注释里直接生成的，比如：

```
// An Example is a mostly-normalized data format for storing data for
// training and inference.  It contains a key-value store (features); where
// each key (string) maps to a Feature message (which is oneof packed
BytesList,
// FloatList, or Int64List).  This flexible and compact format allows the
// storage of large amounts of typed data, but requires that the data shape
// and use be determined by the configuration files and parsers that are used
to
// read and write this format.  That is, the Example is mostly *not* a
// self-describing format.  In TensorFlow, Examples are read in row-major
// format, so any configuration that describes data with rank-2 or above
// should keep this in mind.  For example, to store an M x N matrix of Bytes,
// the BytesList must contain M*N bytes, with M rows of N contiguous values
// each.  That is, the BytesList value must store the matrix as:
```

Example 是用于样本训练和推理的存储数据的规范化数据格式。它包含一个键值存储功能，每个键（字符串）映射到 Feature 的结构。Feature 是一个打包的 BytesList、FloatList 或 Int64 的链表。这种灵活紧凑的格式可以存储大量同类型的数据，但要求数据的形状和使用由用于读取和写入此格式的配置文件和解析器来确定。

也就是说，样本不是一种自描述格式。在 TensorFlow 中，样本以行格式的方式读取，因此在配置维度为 2 或更高维度数据的时候要注意这一点。例如，为了存储字节的 $M \times N$ 矩阵，BytesList 必须包含 $M \times N$ 个字节，每个 M 行具有 N 个连续值。

另一个例子是 Graph：

```
import "third_party/tensorflow/core/framework/node_def.proto";
import "third_party/tensorflow/core/framework/function.proto";
import "third_party/tensorflow/core/framework/versions.proto";

// 展示操作图
message GraphDef {
  repeated NodeDef node = 1;
```

```
  VersionDef versions = 4;

  int32 version = 3 [deprecated = true];

  FunctionDefLibrary library = 2;
};
```

图（Graph）是 TensorFlow 里的一个重要的概念，从上面的定义里我们可以看到，所谓的图就是一系列节点（Node）的集合，请注意里面的两个与 version 有关的成员：

```
int32 version = 3 [deprecated = true];
```

这只是一个 32 位整数型成员，而且它已经不能再被使用了（Deprecated），取代它的是 VersionDef，它的编号是 4，而 verison 的编号是 3。通过这种方式，Protobuf 可以很好地解决向后兼容的问题。代码如下：

```
VersionDef versions = 4;
```

Protobuf 还被广泛用于定义 RPC，我们在这里就不多描述了。

6.1.2　FlatBuffers

FlatBuffers 是一个高效的跨平台序列化库，适用于 C++、C＃、C、Go、Java、JavaScript、Lobster、Lua、TypeScript、PHP、Python 和 Rust。它最初是在谷歌创建的，用于游戏开发及其他性能相关的关键应用程序。

FlatBuffers 里的两个关键概念是 Schema 和 Table。

Table（表）的定义：Table 是 FlatBuffers 的基石，因为格式演变对于大多数序列化应用程序而言至关重要。通常，处理格式更改是可以在大多数序列化解决方案的解析过程中透明地完成的，但是 FlatBuffers 在访问之前是不会被解析的。表的访问是通过使用额外的间接访问字段来实现的，具体地讲是通过 vtable 来解决这个问题的。

每个表都带有一个 vtable（可以在具有相同布局的多个表之间共享），并包含存储此特定类型的 vtable 实例的字段信息。vtable 也可能表示该字段不存在（因为此 FlatBuffers 是使用较旧版本的软件编写的，该实例不需要该信息或被视为已弃用），在这种情况下返回默认值。

表拥有很低的内存开销（因为 vtable 很小并且共享）和访问成本（额外的间接），但它提供了很大的灵活性。表甚至可以比等效结构花费更少的内存，因为当字段等于它们的默认值时就不需要存储它们。FlatBuffers 还提供了“裸”结构，它不提供向前/向后兼容性，但存储的开销可以更小（对于不太可能改变的非常小的对象有用，例如坐标对或 RGBA 颜色）。

FlatBuffers 的模式与现有的 Protobuf 的模式非常相似，对于熟悉 C 语言的读者来说，FlatBuffers 的定义通常应该是易读的。和 Protobuf 相比，FlatBuffers 通过以下方式改进.proto 文件提供的功能：

- 对于弃用字段的情况，现在手动进行字段 ID 分配。
- 区分表格和结构。所有表字段都是可选的，所有结构字段都是必需的。
- 使用原生的矢量类型而不是重复的。
- 拥有本机联合类型 union 而不是使用一系列可选字段。
- 能够为所有标量定义默认值，而不必在每次访问时处理其可选性。
- 一种可以统一处理模式和数据定义（JSON 兼容）的解析器。

下面我们看一下，在 TensorFlow Lite 里 schema.fbs 是如何使用 FlatBuffers 的：

```
namespace tflite;

//  对应版本
file_identifier "TFL3";
// 任何已写入文件的文件扩展名
file_extension "tflite";

// 重要提示：必须在末尾添加表、枚举和联合的所有新成员，以确保向后的兼容性


// 存储在张量中的数据类型
enum TensorType : byte {
  FLOAT32 = 0,
  FLOAT16 = 1,
  INT32 = 2,
  UINT8 = 3,
  INT64 = 4,
  STRING = 5,
  BOOL = 6,
  INT16 = 7,
  COMPLEX64 = 8,
```

```
    }

    table Tensor {
      // 张量类型。每个条目的含义是特定于操作员的，但内置操作使用批量大小、高度、宽度、通道
数（TensorFlow 的 NHWC）
      shape:[int];
      type:TensorType;
      // An index that refers to the buffers table at the root of the model. Or,
      // if there is no data buffer associated (i.e. intermediate results), then
      // this is 0 (which refers to an always existent empty buffer).
      //
      // The data_buffer itself is an opaque container, with the assumption that
the
      // target device is little-endian. In addition, all builtin operators assume
      // 内存的顺序是这样的：如果 shape 是[4, 3, 2]，那么索引[i, j, k]映射到数据缓冲区
[i*3*2 + j*2 + k]
      buffer:uint;
      name:string;  // 用于调试和导入 TensorFlow
      quantization:QuantizationParameters;  // 可选
      is_variable:bool = false;
    }

    // 内置运算符列表。内置运算符比自定义运算符稍快
    // 虽然自定义运算符接受包含配置参数的不透明对象，但内置函数有一组预先确定的可接受选项
    enum BuiltinOperator : byte {
    }

    // 内置运算符的选项
    union BuiltinOptions {
    }

    table Operator {
      // 索引到 operator_codes 数组。在这里使用整数可以避免复杂的地图查找
      opcode_index:uint;

      // 可选输入和输出张量用-1 表示
      inputs:[int];
      outputs:[int];

      builtin_options:BuiltinOptions;
      custom_options:[ubyte];
      custom_options_format:CustomOptionsFormat;
```

```
    mutating_variable_inputs:[bool];
  }

  // 根类型，定义子图，通常表示整个模型
  table SubGraph {
    // 子图中使用的所有张量的列表
    tensors:[Tensor];

    // 输入此子图的张量的索引。注意这是子图进行推理的非静态张量列表
    inputs:[int];

    // 此子图输出的张量的索引。注意这是子图的推理
    outputs:[int];


    operators:[Operator];

    // 子图的名称（用于调试）
    name:string;
  }

  // 原始数据缓冲区表（用于常量张量）
  table Buffer {
    data:[ubyte] (force_align: 16);
  }

  table Model {
    // 架构版本
    version:uint;

    // 此模型中使用的所有操作员代码的列表，它需要保持有序，因为运算符在其中携带索引矢量
    operator_codes:[OperatorCode];

    // 模型的所有子图。假定第 0 个是主模型
    subgraphs:[SubGraph];

    // 模型的描述  description:string;

    // 模型缓冲区
    // 注意，此数组的第 0 个条目必须是空缓冲区（sentinel），这是一个约定，因此没有缓冲区
的张量可以提供 0 作为其缓冲区
    buffers:[Buffer];
```

```
  // 有关模型的元数据。间接进入现有缓冲区列表  metadata_buffer:[int];
}

root_type Model;
```

关于 FlatBuffers 的具体用法，我们来看看 TensorFlow Lite 是怎么用的。在 TensorFlow 里，可以参考 tensorflow/lite/model.h 和 tensorflow/lite/model.cc，这两个文件介绍了很多 FlatBuffers 的使用方法。

在 TensorFlow 里主要使用了 VerifyAndBuildFromFile()，Android 的应用把模型的存储路径通过 JNI 传进来，这个函数先读取模型文件，然后返回一个 FlatBuffers 的模型，以下是相关的代码。

```
JNIEXPORT jlong JNICALL
Java_org_tensorflow_lite_NativeInterpreterWrapper_createModel(
    JNIEnv* env, jclass clazz, jstring model_file, jlong error_handle) {
  BufferErrorReporter* error_reporter =
      convertLongToErrorReporter(env, error_handle);
  if (error_reporter == nullptr) return 0;
  const char* path = env->GetStringUTFChars(model_file, nullptr);

  std::unique_ptr<tflite::TfLiteVerifier> verifier;
  verifier.reset(new JNIFlatBufferVerifier());

  auto model = tflite::FlatBufferModel::VerifyAndBuildFromFile(
      path, verifier.get(), error_reporter);
```

另外一个是 BuildFromBuffer()，与 VerifyAndBuildFromFile()不同的是：模型存储在内存的 Buffer 里，TensorFlow Lite 没有直接使用这个函数。但是如果开发者的应用已经把模型读进了内存，就可以直接调用这个函数。

在这里我们可以特别关注一下 GetAllocationFromFile 函数，有两个理由：第一，这个实现是对应非 MCU 的，或者说是为 Android 定制的。第二，如果 Android 支持 NNAPI，则选择 NNAPI；如果内核支持共享内存 MMAP，则选择使用 MMAP。

```
#ifndef TFLITE_MCU
// 从“文件名”加载模型。如果mmap_file为true，则使用mmap；否则，在缓冲区中复制模型
std::unique_ptr<Allocation> GetAllocationFromFile(const char* filename,
                                                  bool mmap_file,
                                                  ErrorReporter* error_reporter,
                                                  bool use_nnapi) {
```

```
  std::unique_ptr<Allocation> allocation;
  if (mmap_file && MMAPAllocation::IsSupported()) {
    if (use_nnapi && NNAPIDelegate::IsSupported())
      allocation.reset(new NNAPIAllocation(filename, error_reporter));
    else
      allocation.reset(new MMAPAllocation(filename, error_reporter));
  } else {
    allocation.reset(new FileCopyAllocation(filename, error_reporter));
  }
  return allocation;
}
```

当使用 MMAP 的时候，会调用 mmap()，这是 Posix 的 API，这个 API 的实现来自于 Asylo。我们提起 Asylo 的原因是，机器学习的模型都非常大，在移动设备和嵌入式设备上运行的时候，读、写和数据转移都会产生性能上的问题，通过共享内存的方法来解决读写内存的问题是常用的解决方法，但是随之会带来安全上的问题，幸好 Asylo 为我们提供了一个解决方案，并作为 Posix 架构中的一部分。

Asylo 的官方定义是：

Asylo provides strong encapsulation around data and logic for developing and using an enclave. In the Asylo C++ API, an enclave application has trusted and untrusted components. The API has a central manager object for all hosted enclave applications.

Asylo 为开发和使用飞地提供了强大的数据和逻辑封装。在 Asylo C++ API 中，安全区应用程序具有受信任和不受信任的组件。API 具有适用于所有托管安全区应用程序的中央管理器对象。

无论是英特尔的 SGX 还是 ARM TrustZone 都提供了一种隔离机制，用来保护代码和数据免遭修改或泄露。虽然我们使用的很多模型都是开源的，但是越来越多的模型将被开发出来，如果保护模型不被恶意侵犯，这其中既包括保护模型本身，也包括保护模型的运行环境，安全性会变得越来越重要。在国内，机器学习应用的差异化越来越高，应用的范围越来越广，对这方面的重视会与日俱增。在本书中就不着重讨论这点了。

回到 FlatBuffers 本身，在 FlatBuffers 的 build_defs.bzl 里，定义怎样生成 FlatBuffers 的文件，TensorFlow Lite 直接使用了它：

```
# Generic schema for inference on device.
flatbuffer_cc_library(
    name = "schema_fbs",
    srcs = ["schema.fbs"],
```

```
    )
```

当编译 TensorFlow Lite 的时候，schema 会被自动编译，由 FlatBuffers 的编译器在 bazel-out 的文件夹里生成 C 代码和头文件，比如：

```
    bazel-out/android-arm64-v8a-opt/genfiles/tensorflow/lite/schema/schema_g
enerated.h
```

这个文件可以直接被引用，请注意它的前缀路径。

```
    #include "tensorflow/lite/schema/schema_generated.h"
```

在这个头文件中，GetModel 可以根据输入的缓存数据生成一个 FlatBuffers 的 Model，这个函数在 Lite 中被直接使用。关于 FlatBuffers 在 Lite 里的具体应用，我们就不赘述了，大家可以参照源代码。

```
inline const tflite::Model *GetModel(const void *buf) {
  return flatbuffers::GetRoot<tflite::Model>(buf);
}
```

FlatBuffers 的一些优缺点：

FlatBuffers 没有解析步骤，这意味着解析是按需完成的。对于字符串，这可能非常昂贵且缓慢。这在 Android 和移动设备上是非常明显的问题，其中在 UI 线程上的慢操作可能导致丢帧。

FlatBuffers 不提供数据封装，因此不能存储的空间将垃圾回收部分地作为后备缓冲区，这意味着如果 FlatBuffers 中的任何派生对象持续驻留，则整个后备数据阵列也是如此。

FlatBuffers 的核心 schema 会随着程序无限增长，如果将来想要删除一个字段，那是不可能的。

FlatBuffers 的修改更昂贵，需要复制整个后备阵列。

FlatBuffers 的解析和序列化的速度要快得多。

大概总结一下就是，FlatBuffers 生成的代码和库代码都比 Protobuf 小一个数量级或更小，由 FlatBuffers 内置的在线或离线 JSON 解析和生成，是 FlatBuffers 与 gRPC 的无缝集成。

但这并不意味着 FlatBuffers 比 Protobuf 更差，反之亦然。由于设计目标不同导致不同的权衡而设计出不同的数据结构。读者需要根据自己的应用采用更好的数据结构，并且在同一应用程序中，以及在不同的运行环境中使用最合适的数据结构。

6.1.3 模型结构

通常，我们需要先在台式机上设计、训练出目标模型，并将其转化成 TensorFlow Lite 的格式。接着，此格式文件在 TensorFlow Lite 中会被内置 Neon 指令集的解析器加载到内存，并执行相应的计算。由于 TensorFlow Lite 对硬件加速接口有良好的支持，读者可以设计出性能更优的 App 供用户使用。在这里，我们看看 TensorFlow Lite 里的模型文件格式。TensorFlow Lite 定义了 Model 这样的结构体，它是模型的主结构，具体代码如下：

```
table Model {
   version: uint;
   operator_codes: [OperatorCode];
   subgraphs: [SubGraph];

   description: string;
   buffers: [Buffer]
}
```

在上面的 Model 结构体定义中，operator_codes 定义了整个模型的所有算子，subgraphs 定义了所有的子图。在子图当中，第一个元素是主图。buffers 则是数据存储区域，主要存储模型的权重信息。Model 中最重要的部分是 SubGraph，它也是一个结构体：

```
table SubGraph {
   tensors: [Tensor];
   inputs: [int];
   outputs: [int];
   operators: [Operator];

   name: string;
}
```

在 SubGraph 里定义了 Tensor，这也是一个结构体，包含维度、数据类型、Buffer 位置等信息。类似地，tensors 定义了子图的各个 Tensor，而 inputs 和 outputs 用索引的方法维护着子图中 Tensor 与输入输出之间的对应关系。operators 定义了子图当中的算子。

```
table Tensor {
   shape: [int];
   type: TensorType;
   buffer: uint;
```

```
    name: string;
}
```

Buffer 以索引量的形式，给出了这个 Tensor 需要用到子图的哪个 Buffer。

在 SubGraph 中另一个重要的结构体是 Operator，Operator 定义了子图的结构：

```
table Operator {
  opcode_index: uint;
  inputs: [int];
  outputs: [int];
}
```

opcode_index 用索引的方式指明该 Operator 对应了哪个算子。inputs 和 outputs 则是 Tensor 的索引值，指明该 Operator 的输入输出信息。

6.1.4　转换器（Toco）

由于 TensorFlow 使用了新的文件格式和存储模式，TensorFlow Lite 也提供了工具（Toco），读者可以使用它进行模型的转换。这是一个非常重要的工具，它负责把 TensorFlow 的模型转换成 TensorFlow Lite 的模型。下面，让我们来看一下 Toco 是如何工作的。

Toco 的代码位于 tensorflow/lite/toco 文件夹下。Toco 有三个主要功能，即导入、导出和转换。导入将输入转换为 Model 类。导出将模型转换为 tflite 模型或 graphviz。变换基于输入标志对模型进行操作，并且它会删除未使用的算子等。下面是 Toco 的入口函数。

```
    void ToolMain(const ParsedTocoFlags& parsed_toco_flags,
                  const ParsedModelFlags& parsed_model_flags) {
                  ModelFlags model_flags;
                  ReadModelFlagsFromCommandLineFlags(parsed_model_flags,
&model_flags);

     TocoFlags toco_flags;
     ReadTocoFlagsFromCommandLineFlags(parsed_toco_flags, &toco_flags);

     string graph_def_contents;
     ReadInputData(parsed_toco_flags, parsed_model_flags, &toco_flags,
                &model_flags, &graph_def_contents);
                CheckOutputFilePermissions(parsed_toco_flags.output_file);
```

```
  std::unique_ptr<Model> model =
      Import(toco_flags, model_flags, graph_def_contents);
  Transform(toco_flags, model.get());
  string output_file_contents;
  Export(toco_flags, *model, toco_flags.allow_custom_ops(),
      &output_file_contents);
  CHECK(port::file::SetContents(parsed_toco_flags.output_file.value(),
                                output_file_contents, port::file::Defaults())
                                .ok());
}
```

Toco 要做的事其实很简单，就是把模型读入然后输出需要的模型。不过，如果我们了解了这个过程，就会对 TensorFlow Lite 的工作过程有更深入的理解。下面我们来看一下具体实现。

1. 模型读入

它基本上读取 TensorFlow GraphDef 或 TensorFlow Lite 模型，然后转换为张量流模型。分析代码时，一个很有用的工具是 Kythe，使用它可以很快理解程序，也可以快速跳转，笔者一般会把 TensorFlow 重新编译，便于浏览代码，如图 6-2 所示。Kythe 比一般 IDE 内置的代码索引的准确率要高，推荐给有兴趣的读者尝试。

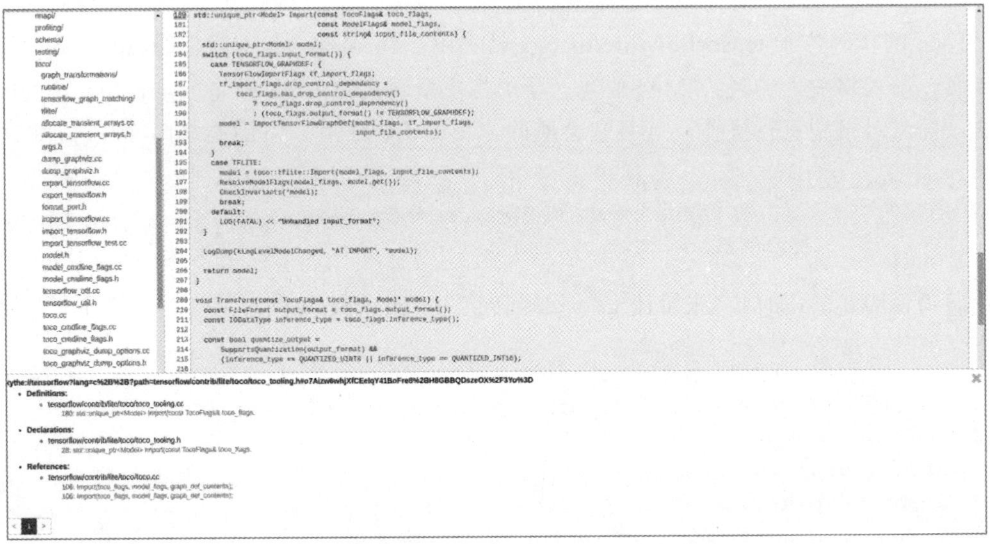

图 6-2　Kythe 代码解析图

比如下面的代码：

```
const ::tflite::Model* input_model =
        ::tflite::GetModel(input_file_contents.data());
```

你可以单击 Model，就会跳转到对应的 FlatBuffers 生成的代码中，这在很多 IDE 里是做不到的，生成的代码如图 6-3 所示。

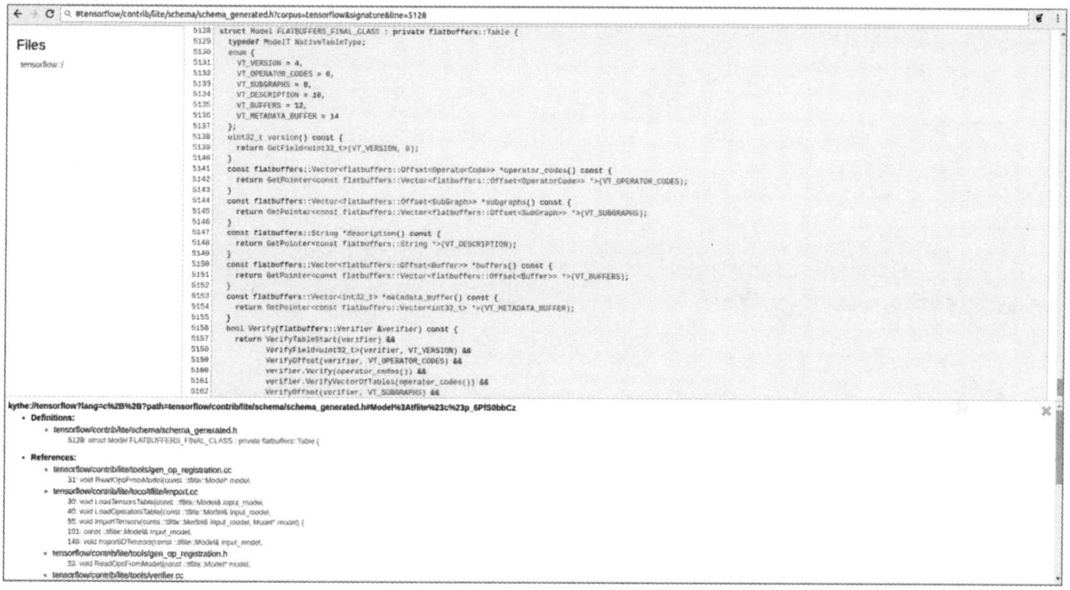

图 6-3　生成代码图

请不要和另一个 model.cc 混淆，它定义了一些辅助函数来复制模型。

现在我们回到 Toco。Toco 主要由两个类来实现。一个类是对应 TensorFlow Lite 的，它是 FlatBuffers 的表示，另一个类的代码在 Toco 文件夹中，它实现了 Toco 的内部逻辑。让我们来说明 Toco 如何将张量流 graphdef 转换为 TensorFlow lite 模型，Toco 主要依靠下面几个函数读取原始模型文件：

```
    ImportTensors(*input_model, model.get());
    ImportOperators(*input_model,        ops_by_name,          tensors_table,
operators_table,
                    model.get());
    ImportIOTensors(*input_model, tensors_table, model.get());
    ImportTensorFlowGraphDef(
        const    ModelFlags&    model_flags,    const    TensorFlowImportFlags&
```

```
tf_import_flags,
      const string& input_file_contents);
```

我们首先来看一下主要函数 ImportTensorFlowGraphDef，它会读入一个 GraphDef 文件数据：

```
    std::unique_ptr<Model> ImportTensorFlowGraphDef(
      const   ModelFlags&   model_flags,   const   TensorFlowImportFlags&
tf_import_flags,
      const string& input_file_contents) {
        std::unique_ptr<GraphDef> tf_graph(new GraphDef);
        CHECK(ParseFromStringEitherTextOrBinary(input_file_contents,
tf_graph.get()));

        std::unique_ptr<GraphDef> pruned_graph =
           MaybeReplaceCompositeSubgraph(*tf_graph);
        if (pruned_graph) {
          tf_graph = std::move(pruned_graph);
        }
        return    ImportTensorFlowGraphDef(model_flags,    tf_import_flags,
*tf_graph);
    }
```

下面两行代码可以把输入数据转化成 GraphDef Protobuf：

```
    std::unique_ptr<GraphDef> tf_graph(new GraphDef);
    CHECK(ParseFromStringEitherTextOrBinary(input_file_contents,
tf_graph.get()));
```

ParseFromStringEitherTextOrBinary 在 toco_port.h 中的定义代码如下，它必须解决谷歌内部构建系统和开源 Protobuf 依赖问题。

```
    std::unique_ptr<Model> ImportTensorFlowGraphDef(
      const   ModelFlags&   model_flags,   const   TensorFlowImportFlags&
tf_import_flags,
          const GraphDef& tf_graph);
```

读入一个 GraphDef 后，我们就要进行模式转换了。下面的函数会先创建一个空模型，然后遍历 GraphDef 中的节点。对于每个节点，调用 ImportTensorFlowNode 将张量流节点转换为 model.h 中的 TensorFlow Lite 的算子。

```
      Model* model = new Model;
```

```
    for (auto node : inlined_graph.node()) {
      StripZeroOutputIndexFromInputs(&node);
      auto status = internal::ImportTensorFlowNode(node, tf_import_flags,
model);
      CHECK(status.ok()) << status.error_message();
    }

    ResolveModelFlags(model_flags, model);

    StripCaretFromArrayNames(model);
    AddExtraOutputs(model);
    FixNoMissingArray(model);
    FixNoOrphanedArray(model);
    FixOperatorOrdering(model);
    CheckInvariants(*model);
```

2. 模型输出

上一节我们介绍了如何将模型转换为 TensorFlow Lite 的模式。export_tensorflow.cc 中的 ExportTensorFlowGraphDef 函数可以输出模型的信息和结构，其定义如下：

```
void ExportTensorFlowGraphDef(const Model& model,
                              string* output_file_contents) {
  CHECK(output_file_contents->empty());
  GraphDef tensorflow_graph;
  ExportTensorFlowGraphDefImplementation(model, &tensorflow_graph);
  LogDumpGraphDef(kLogLevelModelChanged, "AT EXPORT", tensorflow_graph);
  CHECK(tensorflow_graph.SerializeToString(output_file_contents));
```

有兴趣的读者可以从这个函数开始，把实现的内容读清楚。上面我们已经讲解了如何读取 FlatBuffers 文件，TensorFlow Lite 的模型是存储在 FlatBuffers 中的，这里不再赘述。

3. 冻结模型

这里我们讲解一下冻结模型（freeze_graph）的主要功能和它的工作原理。训练后的 TensorFlow 的模型不能被 TensorFlow Lite 直接使用，一定要把它固化（Freeze）。固化的实质是把模型的参数也一同写进同一个模型文件中。它的核心部分是 convert_variables_to_constants，在 python/framework/graph_util_impl.py 中定义如下：

```
def convert_variables_to_constants(sess,
                                   input_graph_def,
```

```
                            output_node_names,
                            variable_names_whitelist=None,
                            variable_names_blacklist=None):
```

本质上，该函数将图形和输出节点名称作为输入参数，从节点跟踪找出子图形，然后在该子图形中用常量替换所有变量。如图 6-4 所示解释了 freeze_graph 中的基本逻辑。

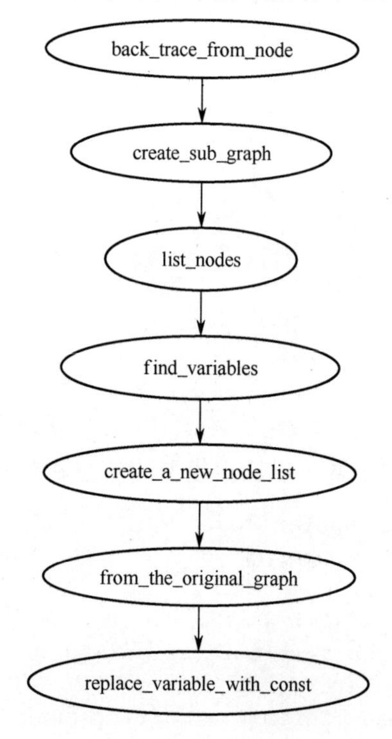

图 6-4　freeze_graph 架构图

下面的代码说明了具体实现的细节：

```
for input_node in inference_graph.node:
  output_node = node_def_pb2.NodeDef()
  if input_node.name in found_variables:
    output_node.op = "Const"
    output_node.name = input_node.name
    dtype = input_node.attr["dtype"]
    data = found_variables[input_node.name]
    output_node.attr["dtype"].CopyFrom(dtype)
    output_node.attr["value"].CopyFrom(
        attr_value_pb2.AttrValue(
```

```
                tensor=tensor_util.make_tensor_proto(
                    data, dtype=dtype.type, shape=data.shape)))
          how_many_converted += 1
        elif input_node.op == "ReadVariableOp" and (
            input_node.input[0] in found_variables):
          # The preceding branch converts all VarHandleOps of ResourceVariables to
          # constants, so we need to convert the associated ReadVariableOps to
          # Identity ops.
          output_node.op = "Identity"
          output_node.name = input_node.name
          output_node.input.extend([input_node.input[0]])
          output_node.attr["T"].CopyFrom(input_node.attr["dtype"])
          if "_class" in input_node.attr:
            output_node.attr["_class"].CopyFrom(input_node.attr["_class"])
        else:
          output_node.CopyFrom(input_node)
        output_graph_def.node.extend([output_node])
```

6.1.5 解析器（Interpreter）

那么 TensorFlow Lite 的解析器又是如何工作的呢？我们来学习一下。

一开始，终端设备会通过 mmap 以内存映射的形式将模型文件载入客户端内存中，其中包含了所有的 Tensor、Operator 和 Buffer 等信息。出于数据使用的需要，TensorFlow Lite 会同时创建 Buffer 的只读区域和分配可写 Buffer 区域。由于解析器中包含了全部执行计算的代码，这一部分被称为 Kernel。模型中的各个 Tensor 会被加载为 TfLiteTensor 的格式并集中存放在 TfLiteContext 中。每个 Tensor 的指针都指向内存中的只读 Buffer 区域，或是一开始新分配的可写入 Buffer 区域。模型中的 Operator 被重新加载为 TfLiteNode，它包含输入输出的 Tensor 索引值。这些 Node 对应的操作符存储于 TfLiteRegistration 中，它包含了指向 Kernel 的函数指针。OpResolver 负责维护函数指针的对应关系。

TensorFlow Lite 在加载模型的过程中会确定执行 Node 的顺序，然后依次执行。

读者如果想要更好地掌握 TensorFlow Lite 的技术细节，一定要阅读以下文件：

```
lite/context.h
lite/model.h
lite/interpreter.h
lite/kernels/register.h
```

程序执行的顺序如表 6-1 所示。

表 6-1

程　　序	文件路径
run	tensorflow/java/src/main/java/org/tensorflow/Session.java
run	tensorflow/java/src/main/native/session_jni.cc
Session::Run	tensorflow/core/common_runtime/session.cc
run	tensorflow/core/common_runtime/direct_session.cc

DirectSession 是 Session 的子类，定义如下：

```
class DirectSession : public Session;
```

解释器是一个静态变量，该变量在加载 TensorFlow 库时被初始化，在其构造函数中，Session 是通过 SessionFactory::Register 注册的。

```
class DirectSessionRegistrar {
 public:
  DirectSessionRegistrar() {
    SessionFactory::Register("DIRECT_SESSION", new
DirectSessionFactory());
  }
};
static DirectSessionRegistrar registrar;
```

解析器是由原生代码（C++）生成的，下面我们来看一下解释器的内部实现代码：

```
JNIEXPORT jlong JNICALL
Java_org_tensorflow_lite_NativeInterpreterWrapper_createInterpreter(
    JNIEnv* env, jclass clazz, jlong model_handle, jlong error_handle,
    jint num_threads) {
  tflite::FlatBufferModel* model = convertLongToModel(env, model_handle);
  if (model == nullptr) return 0;
  BufferErrorReporter* error_reporter =
      convertLongToErrorReporter(env, error_handle);
  if (error_reporter == nullptr) return 0;
  auto resolver = ::tflite::CreateOpResolver();
  std::unique_ptr<tflite::Interpreter> interpreter;
  TfLiteStatus status = tflite::InterpreterBuilder(*model, *(resolver.
get()))(
      &interpreter, static_cast<int>(num_threads));
```

```
    if (status != kTfLiteOk) {
      throwException(env, kIllegalArgumentException,
                     "Internal error: Cannot create interpreter: %s",
                     error_reporter->CachedErrorMessage());
      return 0;
    }
    // 分配内存
    status = interpreter->AllocateTensors();
    if (status != kTfLiteOk) {
      throwException(env, kNullPointerException,
                     "Internal    error:    Cannot    allocate    memory    for    the
interpreter",
                     error_reporter->CachedErrorMessage());
      return 0;
    }
    return reinterpret_cast<jlong>(interpreter.release());
  }
```

在解释器实例的生成过程中包括了所有算子实例的生成，而算子的生成是由重载构成的，对应的代码如下：

```
  TfLiteStatus InterpreterBuilder::operator()(
      std::unique_ptr<Interpreter>* interpreter, int num_threads) {
    if (!interpreter) {
      error_reporter_->Report(
          "Null output pointer passed to InterpreterBuilder.");
      return kTfLiteError;
    }

    // 通过删除部分解释器安全退出，以减少冗长的内容
    auto cleanup_and_error = [&interpreter]() {
      interpreter->reset();
      return kTfLiteError;
    };

    if (!model_) {
      error_reporter_->Report("Null pointer passed in as model.");
      return cleanup_and_error();
    }

    if (model_->version() != TFLITE_SCHEMA_VERSION) {
      error_reporter_->Report(
```

```
          "Model provided is schema version %d not equal "
          "to supported version %d.\n",
          model_->version(), TFLITE_SCHEMA_VERSION);
      return cleanup_and_error();
    }

    if (BuildLocalIndexToRegistrationMapping() != kTfLiteOk) {
      error_reporter_->Report("Registration failed.\n");
      return cleanup_and_error();
    }

    // FlatBuffers 模型的模式定义独立于图形的操作码列表。我们将这些映射到注册表，以便减
少自定义的字符串查找操作，我们只对每个自定义操作执行一次
    auto* subgraphs = model_->subgraphs();
    auto* buffers = model_->buffers();
    if (subgraphs->size() != 1) {
      error_reporter_->Report("Only 1 subgraph is currently supported.\n");
      return cleanup_and_error();
    }
    const tflite::SubGraph* subgraph = (*subgraphs)[0];
    auto operators = subgraph->operators();
    auto tensors = subgraph->tensors();
    if (!operators || !tensors || !buffers) {
      error_reporter_->Report(
          "Did not get operators, tensors, or buffers in input flat buffer.\n");
      return cleanup_and_error();
    }
    interpreter->reset(new Interpreter(error_reporter_));
    if ((**interpreter).AddTensors(tensors->Length()) != kTfLiteOk) {
      return cleanup_and_error();
    }
    // 设置 num 线程
    (**interpreter).SetNumThreads(num_threads);
    // 分析输入/输出
    (**interpreter).SetInputs(FlatBufferIntArrayToVector(subgraph->
inputs()));
    (**interpreter).SetOutputs(FlatBufferIntArrayToVector(subgraph->
outputs()));

    // 最后设置节点和张量
    if (ParseNodes(operators, interpreter->get()) != kTfLiteOk)
      return cleanup_and_error();
```

```
  if (ParseTensors(buffers, tensors, interpreter->get()) != kTfLiteOk)
    return cleanup_and_error();

  return kTfLiteOk;
}
```

解释器生成之后即可启用了：

```
if (interpreter->Invoke() != kTfLiteOk)
```

6.2 底层结构和设计

TensorFlow Lite 提供了 C++和 Java API，在这两套 API 里，API 设计都特别强调了易用性。TensorFlow Lite 是专为移动设备和小型设备上的快速推理而设计的，TensorFlow Lite 希望为开发人员提供简单而高效的 C++ API，以便在 TensorFlow Lite 模型上高效运行推理计算。

6.2.1 设计目标

TensorFlow Lite 是一种推理引擎（Inference Engine），可用于在包括移动设备在内的小型设备上运行。开发者要将 SavedModel 或冻结后的 GraphDef 转换为 TensorFlow Lite 自己的 FlatBuffers 格式。

在使用 TensorFlow Lite 前，开发者需要一种简单的方法来加载他们的模型，将数据提供给推理引擎并得到结果。TensorFlow 的很多代码可以直接在移动设备上运行，但是 TensorFlow Lite 简化了开发者的工作流程，并且能和 TensorFlow 及其他开源架构直接进行协调和工作。

应该注意的是，TensorFlow Lite 通常针对小型设备，而不仅仅是移动设备。这里很难定义小型设备，可以基本认为 TensorFlow Lite 都可以适用于不在数据中心和作为桌面计算设备以外的任何场景。TensorFlow Lite 本身的文件占用空间很小，运行占用的空间也比较小。另外，TensorFlow Lite 对外部库的依赖性也很小。

为了在 TensorFlow Lite 中运行推理模型，用户需要将模型加载到 FlatBufferModel 对象中，然后由 Interpreter 执行。FlatBufferModel 需要保持在 Interpreter 的整个生命周期内有效，并且单独的 FlatBufferModel 可以由多个 Interpreter 同时使用。具体而言，必须在使用它的任何解释器对象之前创建 FlatBufferModel 对象，并且不释放该对象，直到它们全部被销毁。

TensorFlow Lite 的最简单的用法如下所示：

```
    std::unique_ptr<tflite::FlatBufferModel>                model                =
tflite::FlatBufferModel::BuildFromFile(path_to_model);

    tflite::ops::builtin::BuiltinOpResolver resolver;
    tflite::Interpreter   interpreter   =   tflite::Interpreter::Create(model,
resolver);
    interpreter->SetInput(0, std::vector<float>({...}));
    interpreter->Invoke();   const    Tensor*    output    =    interpreter->
GetOutputTensor(0);
```

在上面的代码片段中，我们已经体会到一些 TensorFlow Lite 的设计概要：

- Ops 和 kernels 之间的连接由 OpResolver 提供，如果需要可以使用它对 Ops 进行置换，允许开发者只包含他们所需要的内核。TensorFlow Lite 有自己的内核，这个核心的依赖很少，因此开发人员可以使用自己的 OpResolver 创建非常小的可执行文件。
- 输入和输出张量（Tensor）由整数表示，而不是字符串。在大多数情况下，这是为了避免处理字符串操作，因此代码占用空间很小。通常，使用 TensorFlow Lite 的开发人员已经离线处理了他们的模型，并且确切地知道它们的输入和输出张量的顺序。对于不关心字符串处理代码的读者，可以使用其名称查找到相对应的张量。
- 无法只运行图形的一部分。出于效率原因，TensorFlow Lite 假定图形是按执行顺序定义的，并且所有节点对于推理都是绝对必要的。
- 通过 Tensor 提供对输出张量的访问。在代码内部，TensorFlow Lite 张量是纯 C 结构的，因此内核实现者可以在不依赖 C++的情况下实现操作。TensorFlow Lite 简化了对内部张量数据相关部分的访问。

6.2.2 错误反馈

为了保持较小的二进制大小，TensorFlow Lite 避免依赖于 std::stream 和高级字符串库。在许多地方，TensorFlow Lite 通过简单的 TfLiteStatus 对象返回状态信息，对应的代码如下：

```
typedef enum {
    kTfLiteOk = 0, kTfLiteError = 1 } TfLiteStatus;
    Failures can be easily verified:
    if (status != kTfLiteOk) { // 此处处理错误
    }
```

但进一步的报错需要更多的处理。在已部署的应用程序中，我们不希望触发任何错误处理代码，但是用户仍需要查看错误消息以进行调试，TensorFlow Lite 为详细的错误报告提供了类似 printf 的界面，相关代码如下：

```
class ErrorReporter {
    virtual int Report(const char* format, va_list args) = 0;
};
```

为了向 stderr 报告错误，TensorFlow Lite 提供了 DefaultErrorReporter，开发者可以从 ErrorReporter 派生出来。

6.2.3　装载模型

FlatBufferModel 类封装了一个模型，可以根据模型的存储类型和位置，采用灵活的方式进行构建，具体实现代码如下：

```
class FlatBufferModel {
    // 基于文件构建模型。如果失败，返回 nullptr
static std::unique_ptr<FlatBufferModel> BuildFromFile(
    const char* filename, ErrorReporter* error_reporter);
    // 基于预加载的 FlatBuffer 构建模型。调用方保留缓冲区的所有权，并应保持缓冲区的活
动状态，直到返回的对象被销毁。如果失败，返回 nullptr
static std::unique_ptr<FlatBufferModel> BuildFromBuffer(
    const char* buffer, size_t buffer_size, ErrorReporter* error_reporter);
    // 基于现有的文件描述符构建模型。如果失败则返回 nullptr
 static std::unique_ptr<FlatBufferModel> BuildFromFileDescriptor(
    int    file_descriptor,    size_t    buffer_size,    ErrorReporter*
error_reporter);
};
```

当模型加载文件时，如果 TensorFlow Lite 检测到 Android NNAPI，它将自动尝试使用

共享内存来存储 FlatBufferModel。确保 NNAPI 可用的用户可以传递表示共享内存的文件描述符。Java 用户很可能将他们的模型加载到字节缓冲区并将它们直接传递给 C++ API。

6.2.4 运行模型

运行模型需要几个简单的步骤：

- 基于现有的 FlatBufferModel 构建解释器。
- 选择调整输入张量的大小和形状。出于效率原因，TensorFlow Lite 模型通常已经包含所有张量的大小和形状。
- 设置输入张量值，方法是将数据复制到解释器中或引用外部分配的数据。
- 调用推理。
- 读取输出张量值。
- 可选择重复最后三个步骤：设置输入、调用、读取输出。

解释器（Interpreter）的公共接口开发代码如下：

```
    class Interpreter {

        static TfLiteStatus Create(const FlatBufferModel* model,
        const OpResolver& op_resolver, ErrorReporter* error_reporter);

        // 对输入索引列表的只读访问
        const std::vector<size_t>& inputs() const;

        // 对输出索引列表的只读访问
        const std::vector<size_t>& outputs() const;

        // 改变给定张量的维数。“张量索引”应该是 inputs()返回的索引之一
        TfLiteStatus    ResizeInputTensor(size_t    tensor_index,    const
std::vector<int>& dims);

        // 将数据复制到输入张量中
        template <typename T> TfLiteStatus SetInputTensor(size_t tensor_index,
std::vector<T> data);

        // 将输入张量的值设置为对外部分配的引用内存
```

```
        template <typename T> TfLiteStatus SetInputTensorFromMemory(size_t
tensor_index, T* buffer, size_t buffer_size_in_bytes);

        // 表示外部分配的内存缓冲区的半不透明类型。这在尝试使用 NNAPI 缓冲区设置张量数据时
特别有用
        class MemoryBuffer {

        enum Type {
           ANDROID_SHARED_MEMORY = 0;
        };

        virtual Type GetType() = 0;

        // 将数据复制到给定的缓冲区中。如果复制失败，则返回 false
        virtual bool CopyTo(void* buffer, size_t buffer_size_in_bytes) = 0;

        // 返回内存缓冲区中的字节数
        virtual size_t Size() = 0;
        };

        // 设置输入张量的值作为对内存缓冲区的引用。当后端执行 Ops 时，可以将缓冲区类型强制
转换成内部内存类型，不进行复制
        TfLiteStatus SetInputTensorFromBuffer(size_t tensor_index, MemoryBuffer*
buffer);

        // 返回一个指针，该指针指向存储在输出张量中的值
        // 如果 tensor_index 超出范围或不是输出张量，则返回空指针
        // 对 getOutputtensor()的调用和返回的张量仅在 invoke()之后有效
        const Tensor* GetOutputTensor(size_t tensor_index);

        // 返回 memorybuffer
        MemoryBuffer GetOutputTensor(size_t tensor_ index);

        // 执行模型，填充输出张量。这将使以前从 getOutputEnsor () 获取的引用无效
        // 从 GetOutputTensor () 获取的引用
        TfLiteStatus Invoke();
        };
```

解释器具有如下特性：

- 张量由整数表示，以避免字符串比较，以及对字符串库的任何固定依赖。
- 可以设置输入值而不进行复制。

- 访问输出数据不会暴露内部 TensorFlow Lite 的张量表示，而是返回包装器对象。
- 不能从并发线程访问解释器。
- 在调整张量大小后立即分配内部和输出张量的内存。

6.2.5 定制演算子（CUSTOM Ops）

除了 TensorFlow Lite 自带的演算子，读者还可以定制演算子。由于现在 TensorFlow Lite 自带的演算子非常有限，如果模型变得越来越复杂，可能一段时间内，读者要定制很多演算子。

下面我们来看一下 TensorFlow Lite 是怎样实现的。

TensorFlow Lite 的目标之一是为人们提供构建解释器的基本部分，在这种开发环境里，没有完整的 C++工具链，如果读者的构建目标是 DSP 或其他类型的设备，只能使用 C。TensorFlow Lite 框架提供了几个可用于编写自定义操作的 C 结构：

- TfLiteContext 提供对解释器状态的访问，可用于检索全局对象，包括张量。更常见的是，TfLiteContext 用于报告操作员处理中的错误。
- TfLiteNode 包含有关正在执行的操作的信息。实现可以使用此对象访问其输入和输出张量。

下面是 TfLiteContext 和 TfLiteNode 的定义，代码如下：

```
    struct TfLiteContext {

        // 模型中张量的数目： int tensors_size;
        // 模型中张量的列表： TfLiteTensor* tensors;
        // 更新张量的维数
        TfLiteStatus   (*ResizeTensor)(struct   TfLiteContext*,   TfLiteTensor*
tensor,

        TfLiteIntArray* new_size);
        // 请求用格式字符串 msg 报告错误
        void (*ReportError)(struct TfLiteContext*, const char* msg, ...);

        // 添加 tensors_to_add 张量，保留现有的张量。如果 tensors_to_add 的值为非空，则
First_New_Tensor_Index 指向的值将设置为第一个新张量的索引
```

```
    TfLiteStatus  (*AddTensors)(struct  TfLiteContext*,  size_t  tensors_
to_add,

    int* first_new_tensor_index); };

    struct TfLiteNode {

    // 此节点的输入表示为 TfLiteContext 张量的索引
    TfLiteIntArray* inputs;

    // 输出到该节点，表示为 TfLiteContext 张量的索引
    TfLiteIntArray* outputs;

    // 该节点的 init()函数返回的不透明数据（见下文）
    void* user_data;

    // 在输入 FlatBuffers 时为该节点提供的不透明数据。这只适用于内置操作
    void* builtin_data;
  };
```

编写 TensorFlow Lite 内核涉及定义四个 C 函数：init()、prepare()、invoke()和 free()，无论是自定义还是内置操作都是同样的，相关代码如下：

```
  typedef struct _TfLiteRegistration {

  // 初始化序列化数据中的操作
  // 如果是内置 OP：buffer 操作的参数数据（tflitelstmparams*）的长度为零
  // 如果是自定义 OP：buffer 是操作的“自定义”选项
  // length 是缓冲区的大小
  // 返回类型 punned（即 void*）是不透明数据（如基元指针或结构的实例）
  // 返回的指针将与节点一起存储在 user_data 字段中，可在下面的准备和调用函数中访问
  // 注意：如果数据已经是所需的格式，只需实现返回 nullptr 的函数，并将自由函数实现为 no-op
   void* (*init)(TfLiteContext* context, const char* buffer, size_t length);

   // 指针 buffer 是先前由 init 调用返回的数据
   void (*free)(TfLiteContext* context, void* buffer);

   // 当此节点所依赖的输入已调整大小时，将调用 Prepare
   // 可以通过调用 context->resizetensor()来请求调整输出张量的大小

   // 成功返回 ktfliteok
   TfLiteStatus (*prepare)(TfLiteContext* context, TfLiteNode* node);
```

```
  // 执行节点 (应该取 node->inputs 和 output 到 node->outputs)
  // 成功返回 kTfLiteOk
  TfLiteStatus (*invoke)(TfLiteContext* context, TfLiteNode* node);

  // 在概要分析信息期间调用概要分析字符串，以便使执行分组在一起
  const char* (*profiling_string)(const TfLiteContext* context,
                                  const TfLiteNode* node);


// 内置代码。注意：注册绑定器负责设置要正确
  int32_t builtin_code;

  // Custom op name. If the op is a builtin, this will be null.
  // Note: It is the responsibility of the registration binder to set this
  // properly.
  // WARNING: This is an experimental interface that is subject to change.
// 自定义操作名称。如果 OP 是内置的，那么它将为空
// 注意：要正确设置注册绑定器
// 注意：这是一个可更改的实验接口
  const char* custom_name;

  // 操作的版本
  int version;
} TfLiteRegistration;
```

当解释器加载模型时，它会为图中的每个节点调用一次 init()。这意味着如果在图中多次使用给定的 init()将被多次调用。init()的声明方式如下：

```
void* (*init)(TfLiteContext* context, const char* buffer, size_t length);
```

使用 FlexBuffers 的注意事项：自定义操作的缓冲区不必是 FlexBuffer。我们将使用枚举来定义每个自定义缓冲区，该枚举描述它包含的数据类型：

```
enum CustomOpDataType {
   FLEXBUFFER = 0;
};
```

如果我们愿意，可以在未来使用不同的类型数据作为缓冲。对于每个init()调用，都会有一个相对应的 free()的调用，允许实现处理它们在调用 init()时返回的缓冲区：

```
void (*free)(TfLiteContext* context, void* init_data);
```

每当输入张量调整大小时，解释器将通过图表通知变更。这使他们有机会调整内部缓冲区的大小，检查输入形状和类型的有效性，并重新计算输出形状。处理此阶段的函数代码如下：

```
TfLiteStatus (*prepare)(TfLiteContext* context, TfLiteNode* node);
```

另外，每次推理运行时，解释器会遍历图形且调用 invoke()：

```
TfLiteStatus (*invoke)(TfLiteContext* context, TfLiteNode* node);
```

custom 和 builtin Ops 都提供了全局注册功能，定义如下所示：

```
namespace tflite {
    namespace ops {
        namespace builtin {
            TfLiteRegistration* Register_MY_CUSTOM_OP() {
                static  TfLiteRegistration  r  =  {  my_custom_op::Init,
my_custom_op::Free, my_custom_op::Prepare, my_custom_op::Eval};
                return &r;
            }
        } // namespace builtin
    } // namespace Ops
} // namespace tflite
```

请注意，注册不是自动的，应该在某处显式调用 Register_MY_CUSTOM_OP。

比如下面的代码要注册一个叫 MY_CUSTOM_OP 的算子：

```
namespace tflite {
namespace ops {
namespace custom {
  TfLiteRegistration* Register_MY_CUSTOM_OP() {
    static TfLiteRegistration r = {my_custom_op::Init,
                                   my_custom_op::Free,
                                   my_custom_op::Prepare,
                                   my_custom_op::Eval};
    return &r;
  }
}  // namespace custom
}  // namespace Ops
}  // namespace tflite
```

然后，调用注册函数。TensorFlow Lite 是在 BuiltinOpResolver::BuiltinOpResolver()中注册运算子的，比如下面的代码注册了五个定制的算子：

```
    AddCustom("Mfcc", tflite::ops::custom::Register_MFCC());
    AddCustom("AudioSpectrogram",
            tflite::ops::custom::Register_AUDIO_SPECTROGRAM());
    AddCustom("LayerNormLstm",
tflite::ops::custom::Register_LAYER_NORM_LSTM());
    AddCustom("Relu1", tflite::ops::custom::Register_RELU_1());
    AddCustom("TFLite_Detection_PostProcess",
            tflite::ops::custom::Register_DETECTION_POSTPROCESS());
```

这样我们就完成了注册定制算子的过程。

6.2.6 定制内核

实际实现时，解释器将加载一个内核库，这些内核将被分配用于执行模型中的每个操作符。虽然默认库仅包含内置内核，但可以使用自定义库替换它。

解释器使用 OpResolver 将操作代码和名称转换为实际代码：

```
    class OpResolver {

      virtual TfLiteRegistration* FindOp(tflite::BuiltinOperator op) const = 0;
      virtual TfLiteRegistration* FindOp(const char* op) const = 0;
      virtual void AddOp(tflite::BuiltinOperator op,

      TfLiteRegistration* registration) = 0;
      virtual void AddOp(const char* op, TfLiteRegistration* registration) = 0;
    };
```

最常见的用法仅依赖于 BuiltinOperator 枚举定义操作名称，避免使用'const char *。在某些应用程序中，避免字符串比较和 std::set 很重要。自定义内核必然会产生字符串查找，如果这些内核的数量很少，查找速度也会很快。在任何情况下，在解释器初始化时，每个节点都会发生一次查找。

如果开发人员希望在现有的解析器中添加一个或两个自定义操作。他们可以使用 BuiltinOpResolver 并手动注册自定义操作，如下所示：

```
    tflite::ops::builtin::BuiltinOpResolver resolver;
    TfLiteStatus   status   =   resolver.AddOp("my_custom_op",   Register_MY_
CUSTOM_OP());
```

注册已存在的操作是错误的。如果内置操作集被认为太大，则可以基于给定的操作子集用代码生成新的 OpResolver。

6.3　工具

由于 TensorFlow Lite 和其他 TensorFlow 组件比较起来比较新，有几个工具非常有用，可以帮助读者了解 TensorFlow Lite，下面我们介绍两个有用的工具。

6.3.1　图像标注（label_image）

这个工具在 tensorflow/lite/examples/label_image 里，它是一个用 C++写的图像分类工具。它会读入一个图像和模型文件，按照标注的文件进行分类。

首先，下载图像标注要用到的几个文件，代码如下：

```
$ ./tensorflow/lite/examples/ios/download_models.sh
download_models.sh 文件的功能是下载以下几个文件：
./tensorflow/lite/examples/ios/simple/data/mobilenet_v1_1.0_224.tflite
./tensorflow/lite/examples/ios/camera/data/mobilenet_quant_v1_224.tflite
./tensorflow/lite/examples/ios/camera/data/mobilenet_v1_1.0_224.tflite
```

然后，构建执行文件。Android 设备的 ABI 可以通过以下命令得到：

```
adb shell getprop | grep abi
```

假设你的设备是 armv8 64 位芯片，可以执行下面的命令：

```
$ bazel build --config android_arm64 --config monolithic --cxxopt=-std=c++11 \
  //tensorflow/lite/examples/label_image:label_image
```

注意我们需要--config monolithic，不然会出现编译问题。

编译后会得到 bazel-bin/tensorflow/lite/examples/label_image/label_image，然后我们把它和必要的文件直接拷贝到一台 Android 手机上并执行下面的命令：

```
# 复制执行文件
$  adb  push  bazel-bin/tensorflow/lite/examples/label_image/label_image
/data/local/tmp
bazel-bin/tensorflow/lite/examples/label_image/label_image: 1 file pushed.
19.5 MB/s (1751824 bytes in 0.086s)

# 复制模型文件
```

```
    $ adb push ./tensorflow/lite/examples/ios/camera/data/mobilenet_quant_
v1_224.tflite /data/local/tmp
    ./tensorflow/lite/examples/ios/camera/data/mobilenet_quant_v1_224.tflite:
1 file pushed. 18.9 MB/s (4276100 bytes in 0.216s)

    # 复制测试图像文件
    $ adb push ./tensorflow/lite/examples/label_image/testdata/grace_
hopper.bmp /data/local/tmp
    ./tensorflow/lite/examples/label_image/testdata/grace_hopper.bmp: 1 file
pushed. 13.8 MB/s (940650 bytes in 0.065s)

    # 复制标注结果文件
    $ adb push ./tensorflow/lite/examples/ios/simple/data/labels.txt /data/
local/tmp
    ./tensorflow/lite/examples/ios/simple/data/labels.txt: 1 file pushed. 0.7
MB/s (10484 bytes in 0.014s)
```

下面，在 Android 的手机上执行命令“$adb shell”，把运行环境从桌面跳转到手机，此时可以看到提示符“# taimen”，这是笔者的手机的名称。手机的品牌不同，提示符也会不同。

如果读者是 Android 的开发者，对解锁（Unlock）应该不陌生。由于我们要向手机设备里拷贝文件，需要额外的权限，所以手机需要先解锁。由于手机和运营商不同，解锁的过程不同，读者需要自己去了解解锁过程。

现在，通过执行命令“taimen:/ # cd /data/local/tmp”把当前路径转移到/data/local/tmp 下，因为在这个路径下，我们可以运行 Android 的应用。

执行命令“taimen:/ # ls -all”，得到以下文件列表：

```
    drwxrwx--x 3 shell shell    4096 2019-01-15 01:35:11.651012625 -0500 .
    drwxr-x--x 4 root  root     4096 2019-01-07 18:26:05.842333353 -0500 ..
    drwxrwxrwx 5 shell shell     4096 2019-01-07 19:34:27.222741927 -0500
deployment
    -rw-rw-rw- 1 root  root   940650 2019-01-14 14:24:27.000000000 -0500
grace_hopper.bmp
    -r-xr-xr-x 1 root  root  1751824 2019-01-15 00:56:19.000000000 -0500
label_image
    -rw-rw-rw- 1 root  root    10484 2019-01-15 01:05:16.000000000 -0500
labels.txt
    -rw-rw-rw- 1 root  root  4276100 2019-01-15 01:05:16.000000000 -0500
```

```
mobilenet_quant_v1_224.tflite
```

然后，执行 label_image 脚本来检测图像识别的准确度，运行结果如下：

```
    taimen:/data/local/tmp # ./label_image
    WARNING: linker: Warning: "/data/local/tmp/label_image" unused DT entry:
DT_RPATH (type 0xf arg 0x488) (ignoring)
    Loaded model ./mobilenet_quant_v1_224.tflite
    resolved reporter
    invoked
    average time: 79.017 ms
    0.666667: 458 bow tie
    0.290196: 653 military uniform
    0.0117647: 835 suit
    0.00784314: 611 jersey
    0.00392157: 922 book jacket
```

这些工具非常有用。因为 Android 开发基本都用 Java，所以我们要写 Java 的应用并使其与 JNI 连接起来。可是，TensorFlow 都是通过 C++实现的，在对性能要求高的设备上，用 C++的工具去测试会方便很多，提高工作效率。比如，可以先用这个工具去测试性能，可以省去很多不必要的 Android 操作，而且可以测试更多的场景。我们看到了，这个工具是用 C++写的，那么它经过编译以后，应该可以在各种平台上运行。在以*nix 为基础的平台上，比如 Ubuntu、CentOS、Mac 上也可以执行。

下面来看一下这个工具的一些有用的功能。由于这些功能是共通的，我们也可以在本地机上运行这些功能。在 X86 的 Ubuntu、Mac 或类似的机器上执行如下代码：

```
    $ bazel run --cxxopt=-std=c++11 //tensorflow/lite/examples/label_image -- -h
    label_image
    --accelerated, -a: [0|1], use Android NNAPI or not
    --count, -c: loop interpreter->Invoke() for certain times
    --input_mean, -b: input mean
    --input_std, -s: input standard deviation
    --image, -i: image_name.bmp
    --labels, -l: labels for the model
    --tflite_model, -m: model_name.tflite
    --profiling, -p: [0|1], profiling or not
    --num_results, -r: number of results to show
    --threads, -t: number of threads
    --verbose, -v: [0|1] print more information
```

由于这个工具实际没有-h 帮助选项，我们需要让程序触发错误，然后生成帮助菜单。

我们可以用下面三个选项指定不同的输入文件：

```
    --image, -i: image_name.bmp
    --labels, -l: labels for the model
    --tflite_model, -m: model_name.tflite
```

比如，可以把模型，标注文件和图形文件都保存在本地：

```
    $ bazel-bin/tensorflow/lite/examples/label_image/label_image -i ./grace_
hopper.bmp -l ./labels.txt -m ./mobilenet_quant_v1_224.tflite
```

如果现在使用详细输出选项（Verbose），那么输出结果应该是这样的：

```
    $ bazel-bin/tensorflow/lite/examples/label_image/label_image -i ./grace_
hopper.bmp -l ./labels.txt -m ./mobilenet_quant_v1_224.tflite -v 1
    Loaded model ./mobilenet_quant_v1_224.tflite
    resolved reporter
    tensors size: 89
    nodes size: 31
    inputs: 1
    input(0) name: Placeholder
    0: MobilenetV1/Logits/AvgPool_1a/AvgPool, 1024, 3, 0.0235285, 0
    1: MobilenetV1/Logits/Conv2d_1c_1x1/BiasAdd, 1001, 3, 0.165351, 74
    2: MobilenetV1/Logits/Conv2d_1c_1x1/Conv2D_bias, 4004, 2, 0.000116509, 0
    3:
MobilenetV1/Logits/Conv2d_1c_1x1/weights_quant/FakeQuantWithMinMaxVars,
1025024, 3, 0.00495183, 67
    4: MobilenetV1/MobilenetV1/Conv2d_0/Conv2D_Fold_bias, 128, 2, 0.000161006, 0
    5: MobilenetV1/MobilenetV1/Conv2d_0/Relu6, 401408, 3, 0.0235285, 0
    6:
MobilenetV1/MobilenetV1/Conv2d_0/weights_quant/FakeQuantWithMinMaxVars, 864,
3, 0.0410565, 108
    7: MobilenetV1/MobilenetV1/Conv2d_10_depthwise/Relu6, 100352, 3, 0.0235285, 0
    8: MobilenetV1/MobilenetV1/Conv2d_10_depthwise/depthwise_Fold_bias, 2048,
2, 0.00039105, 0
    9: MobilenetV1/MobilenetV1/Conv2d_10_depthwise/weights_quant/
FakeQuantWithMinMaxVars, 4608, 3, 0.0166203, 131
    此处省略中间结果
    80: MobilenetV1/MobilenetV1/Conv2d_9_depthwise/depthwise_Fold_bias, 2048,
2, 0.000351091, 0
    81: MobilenetV1/MobilenetV1/Conv2d_9_depthwise/weights_quant/
FakeQuantWithMinMaxVars, 4608, 3, 0.014922, 132
    82: MobilenetV1/MobilenetV1/Conv2d_9_pointwise/Conv2D_Fold_bias, 2048, 2,
```

```
0.000161186, 0
    83: MobilenetV1/MobilenetV1/Conv2d_9_pointwise/Relu6, 100352, 3, 0.0235285, 0
    84: MobilenetV1/MobilenetV1/Conv2d_9_pointwise/weights_quant/
FakeQuantWithMinMaxVars, 262144, 3, 0.00685069, 120
    85: MobilenetV1/Predictions/Reshape, 1001, 3, 0.165351, 74
    86: MobilenetV1/Predictions/Reshape/shape, 8, 2, 0, 0
    87: MobilenetV1/Predictions/Softmax, 1001, 3, 0.00390625, 0
    88: Placeholder, 150528, 3, 0.00392157, 0
    len: 940650
    width, height, channels: 517, 606, 3
    input: 88
    number of inputs: 1
    number of outputs: 1
    Interpreter has 90 tensors and 31 nodes
    Inputs: 88
    Outputs: 87

    Tensor          0    MobilenetV1/Logits/AvgPool_1a/AvgPool    kTfLiteUInt8
kTfLiteArenaRw      1024 bytes ( 0.0 MB)  1 1 1 1024
    Tensor         1   MobilenetV1/Logits/Conv2d_1c_1x1/BiasAdd   kTfLiteUInt8
kTfLiteArenaRw      1001 bytes ( 0.0 MB)  1 1 1 1001
    Tensor      2  MobilenetV1/Logits/Conv2d_1c_1x1/Conv2D_bias  kTfLiteInt32
kTfLiteMmapRo      4004 bytes ( 0.0 MB)  1001
    Tensor   3 MobilenetV1/Logits/Conv2d_1c_1x1/weights_quant/
FakeQuantWithMinMaxVars kTfLiteUInt8   kTfLiteMmapRo    1025024 bytes ( 1.0 MB)
1001 1 1 1024
    Tensor   4 MobilenetV1/MobilenetV1/Conv2d_0/Conv2D_Fold_bias kTfLiteInt32
kTfLiteMmapRo       128 bytes ( 0.0 MB)  32
    Tensor   5 MobilenetV1/MobilenetV1/Conv2d_0/Relu6 kTfLiteUInt8
kTfLiteArenaRw    401408 bytes ( 0.4 MB)  1 112 112 32
    Tensor   6 MobilenetV1/MobilenetV1/Conv2d_0/weights_quant/
FakeQuantWithMinMaxVars kTfLiteUInt8   kTfLiteMmapRo       864 bytes ( 0.0 MB)
32 3 3 3
    Tensor   7 MobilenetV1/MobilenetV1/Conv2d_10_depthwise/Relu6 kTfLiteUInt8
kTfLiteArenaRw    100352 bytes ( 0.1 MB)  1 14 14 512
    Tensor   8 MobilenetV1/MobilenetV1/Conv2d_10_depthwise/depthwise_
Fold_bias kTfLiteInt32   kTfLiteMmapRo      2048 bytes ( 0.0 MB)  512
    Tensor   9 MobilenetV1/MobilenetV1/Conv2d_10_depthwise/weights_quant/
FakeQuantWithMinMaxVars kTfLiteUInt8   kTfLiteMmapRo      4608 bytes ( 0.0 MB)
1 3 3 512
    Tensor   10  MobilenetV1/MobilenetV1/Conv2d_10_pointwise/Conv2D_Fold_bias
```

```
kTfLiteInt32   kTfLiteMmapRo      2048 bytes ( 0.0 MB)  512
    此处省略中间结果
    Tensor  80 MobilenetV1/MobilenetV1/Conv2d_9_depthwise/depthwise_Fold_bias
kTfLiteInt32   kTfLiteMmapRo      2048 bytes ( 0.0 MB)  512
    Tensor                                                                81
MobilenetV1/MobilenetV1/Conv2d_9_depthwise/weights_quant/FakeQuantWithMinMax
Vars kTfLiteUInt8   kTfLiteMmapRo      4608 bytes ( 0.0 MB)  1 3 3 512
    Tensor   82  MobilenetV1/MobilenetV1/Conv2d_9_pointwise/Conv2D_Fold_bias
kTfLiteInt32   kTfLiteMmapRo      2048 bytes ( 0.0 MB)  512
    Tensor  83 MobilenetV1/MobilenetV1/Conv2d_9_pointwise/Relu6 kTfLiteUInt8
kTfLiteArenaRw    100352 bytes ( 0.1 MB)  1 14 14 512
    Tensor                                                                84
MobilenetV1/MobilenetV1/Conv2d_9_pointwise/weights_quant/FakeQuantWithMinMax
Vars kTfLiteUInt8   kTfLiteMmapRo    262144 bytes ( 0.2 MB)  512 1 1 512
    Tensor  85  MobilenetV1/Predictions/Reshape  kTfLiteUInt8   kTfLiteArenaRw
1001 bytes ( 0.0 MB)  1 1001
    Tensor       86    MobilenetV1/Predictions/Reshape/shape    kTfLiteInt32
kTfLiteMmapRo          8 bytes ( 0.0 MB)  2
    Tensor  87  MobilenetV1/Predictions/Softmax  kTfLiteUInt8   kTfLiteArenaRw
1001 bytes ( 0.0 MB)  1 1001
    Tensor  88  Placeholder            kTfLiteUInt8   kTfLiteArenaRw      150528
bytes ( 0.1 MB)  1 224 224 3
    Tensor  89 (null)              kTfLiteUInt8  kTfLiteArenaRw     338688 bytes
( 0.3 MB)  1 112 112 27

    Node   0 Operator Builtin Code   3
      Inputs: 88 6 4
      Outputs: 5
    Node   1 Operator Builtin Code   4
      Inputs: 5 33 32
      Outputs: 31
    Node   2 Operator Builtin Code   3
      Inputs: 31 36 34
      Outputs: 35
    Node   3 Operator Builtin Code   4
      Inputs: 35 39 38
      Outputs: 37
    Node   4 Operator Builtin Code   3
      Inputs: 37 42 40
      Outputs: 41
    Node   5 Operator Builtin Code   4
```

```
  Inputs: 41 45 44
  Outputs: 43
Node  6 Operator Builtin Code  3
  Inputs: 43 48 46
  Outputs: 47
Node  7 Operator Builtin Code  4
  Inputs: 47 51 50
  Outputs: 49
Node  8 Operator Builtin Code  3
  Inputs: 49 54 52
  Outputs: 53
Node  9 Operator Builtin Code  4
  Inputs: 53 57 56
  Outputs: 55
Node 10 Operator Builtin Code  3
  Inputs: 55 60 58
  Outputs: 59
此处省略中间结果
Node 26 Operator Builtin Code  3
  Inputs: 25 30 28
  Outputs: 29
Node 27 Operator Builtin Code  1
  Inputs: 29
  Outputs: 0
Node 28 Operator Builtin Code  3
  Inputs: 0 3 2
  Outputs: 1
Node 29 Operator Builtin Code 22
  Inputs: 1 86
  Outputs: 85
Node 30 Operator Builtin Code 25
  Inputs: 85
  Outputs: 87
invoked
average time: 203.584 ms
0.666667: 458 bow tie
0.290196: 653 military uniform
0.0117647: 835 suit
0.00784314: 611 jersey
0.00392157: 922 book jacket
```

输出的内容里有一些有意义的信息，比如：

```
# 模型里包含的张量和节点（node）
Interpreter has 90 tensors and 31 nodes
Inputs: 88
Outputs: 87
# 推测所用的时间
average time: 203.584 ms
```

下面我们通过代码来了解这个应用是怎样实现的。这个应用只支持 BMP 图像格式，读取 BMP 是由 read_bmp()实现的。下面的代码是 read_bmp()的一部分，功能是直接从 BMP 文件里读取图像解析度的宽和高。这段代码的优点是，实现比较简单，同时能够避免对第三方代码库的依赖。

```
const int32_t header_size =
    *(reinterpret_cast<const int32_t*>(img_bytes.data() + 10));
  *width = *(reinterpret_cast<const int32_t*>(img_bytes.data() + 18));
  *height = *(reinterpret_cast<const int32_t*>(img_bytes.data() + 22));
  const int32_t bpp =
    *(reinterpret_cast<const int32_t*>(img_bytes.data() + 28));
```

回到 label_image.cc，这里的 void RunInference(Settings* s)实现了 TensorFlow Lite 的推理逻辑。

首先，使用 FlatBuffers 的接口函数从模型的文件名构建一个模型的实例：

```
std::unique_ptr<tflite::FlatBufferModel> model;
model = tflite::FlatBufferModel::BuildFromFile(s->model_name.c_str());
```

然后，调用 TfLiteStatus InterpreterBuilder::operator()()生成一个 tflite::Interpreter 的实例：

```
std::unique_ptr<tflite::Interpreter> interpreter;
tflite::ops::builtin::BuiltinOpResolver resolver;
tflite::InterpreterBuilder(*model, resolver)(&interpreter);
```

接着，为模型分配内存。内存管理基本是由 ArenaPlanner 类来实现的，代码如下：

```
if (interpreter->AllocateTensors() != kTfLiteOk) {
    LOG(FATAL) << "Failed to allocate tensors!";
}
```

最后，调用 Invoke()执行模型的推理：

```
for (int i = 0; i < s->loop_count; i++) {
  if (interpreter->Invoke() != kTfLiteOk) {
    LOG(FATAL) << "Failed to invoke tflite!\n";
  }}
```

上面的代码构成了使用 TensorFlow Lite 的基本架构，应该还是比较简单的。

在这个应用里，有两个小的特点读者可以留意一下。

一个特点是使用了 PrintInterpreterState，使用它可以把 Interpreter 的内部状态打印输出：

```
if (s->verbose) PrintInterpreterState(interpreter.get());
```

它的实现也比较简单，基本的逻辑是找到模型的节点和张量，并输出它们的状态：

```
  for (size_t tensor_index = 0; tensor_index < interpreter->tensors_size();
       tensor_index++) {
    ... ...
  }
  printf("\n");
  for (size_t node_index = 0; node_index < interpreter->nodes_size();
       node_index++) {
    ... ...
  }
```

另外一个特点是使用了 Profiler，通过 Profiler 可以把运行状态打印出来。调用 Profiler 的代码如下：

```
profiling::Profiler* profiler = new profiling::Profiler();
interpreter->SetProfiler(profiler);
if (s->profiling) profiler->StartProfiling();
```

激活 TFLITE_PROFILING_ENABLED，重新编译：

```
$ bazel build --cxxopt=-std=c++11 --copt=-DTFLITE_PROFILING_ENABLED
//tensorflow/lite/examples/label_image
```

执行结果如下：

```
$ bazel-bin/tensorflow/lite/examples/label_image/label_image -i ./grace_
hopper.bmp -l ./labels.txt -m ./mobilenet_quant_v1_224.tflite -p 1
Loaded model ./mobilenet_quant_v1_224.tflite
resolved reporter
invoked
average time: 210.206 ms
```

```
    13.280, Node  0, OpCode  3, CONV_2D
     7.841, Node  1, OpCode  4, DEPTHWISE_CONV_2D
    11.299, Node  2, OpCode  3, CONV_2D
     3.891, Node  3, OpCode  4, DEPTHWISE_CONV_2D
     8.465, Node  4, OpCode  3, CONV_2D
     7.681, Node  5, OpCode  4, DEPTHWISE_CONV_2D
    15.893, Node  6, OpCode  3, CONV_2D
     1.927, Node  7, OpCode  4, DEPTHWISE_CONV_2D
     7.634, Node  8, OpCode  3, CONV_2D
     3.713, Node  9, OpCode  4, DEPTHWISE_CONV_2D
    14.745, Node 10, OpCode  3, CONV_2D
     0.934, Node 11, OpCode  4, DEPTHWISE_CONV_2D
     7.308, Node 12, OpCode  3, CONV_2D
     1.781, Node 13, OpCode  4, DEPTHWISE_CONV_2D
    14.232, Node 14, OpCode  3, CONV_2D
     1.779, Node 15, OpCode  4, DEPTHWISE_CONV_2D
    14.258, Node 16, OpCode  3, CONV_2D
     1.780, Node 17, OpCode  4, DEPTHWISE_CONV_2D
    14.243, Node 18, OpCode  3, CONV_2D
     1.779, Node 19, OpCode  4, DEPTHWISE_CONV_2D
    14.269, Node 20, OpCode  3, CONV_2D
     1.778, Node 21, OpCode  4, DEPTHWISE_CONV_2D
    14.242, Node 22, OpCode  3, CONV_2D
     0.462, Node 23, OpCode  4, DEPTHWISE_CONV_2D
     7.633, Node 24, OpCode  3, CONV_2D
     0.822, Node 25, OpCode  4, DEPTHWISE_CONV_2D
    15.075, Node 26, OpCode  3, CONV_2D
     0.032, Node 27, OpCode  1, AVERAGE_POOL_2D
     1.342, Node 28, OpCode  3, CONV_2D
     0.000, Node 29, OpCode 22, RESHAPE
     0.088, Node 30, OpCode 25, SOFTMAX
0.667: 458 bow tie
0.290: 653 military uniform
0.012: 835 suit
0.008: 611 jersey
0.004: 922 book jacket
```

从结果可以发现卷积 CONV_2D 使用的计算时间最多。Profiler 是由 ProfileBuffer 类实现的，它的基本功能就是记录各个事件的时间点。由于它是内置的功能，因此使用它的好处是，不需要额外写代码来实现这些功能。

6.3.2 最小集成（Minimal）

在 tensorflow/lite/examples/下的 Minimal 是一个由 GitHub 社区贡献的小工具。它演示了怎样读入模型、构建解释器，以及运行预测。Minimal 可以作为其他工具的代码基础，它没有太复杂的代码，只是提供了 build 文件的写法。如果我们构建一个执行文件，只需使用 tf_cc_binary 并替换 srcs 中的源文件，非常简单，代码如下：

```
tf_cc_binary(
    name = "minimal",
    srcs = [
        "minimal.cc",
    ],
    linkopts = tflite_linkopts() + select({
        "//third_party/tensorflow:android": [
            "-pie",  # Android 5.0 and later supports only PIE
            "-lm",  # some builtin ops, e.g., tanh, need -lm
        ],
        "//conditions:default": [],
    }),
    deps = [
        "//third_party/tensorflow/lite:framework",
        "//third_party/tensorflow/lite/kernels:builtin_ops",
    ],
)
```

6.3.3 Graphviz

可视化是机器学习中理解数据的方法，也是理解代码和内部逻辑的重要方法。TensorFlow 提供了很多可视化工具，这里让我们尝试一些有趣的转换，比如生成张量流图。

首先，将 TensorFlow 张量流图转换为 Graphviz：

```
$ toco --input_file=tf_files/retrained_graph.pb --output_file=tf_files/retrained.dot --input_format=TENSORFLOW_GRAPHDEF --output_format=GRAPHVIZ_DOT --input_shape=1,224,224,3  --input_array=input  --output_array=final_result --inference_type=FLOAT --input_data_type=FLOAT
```

Graphviz 生成的模型图如图 6-5 所示。

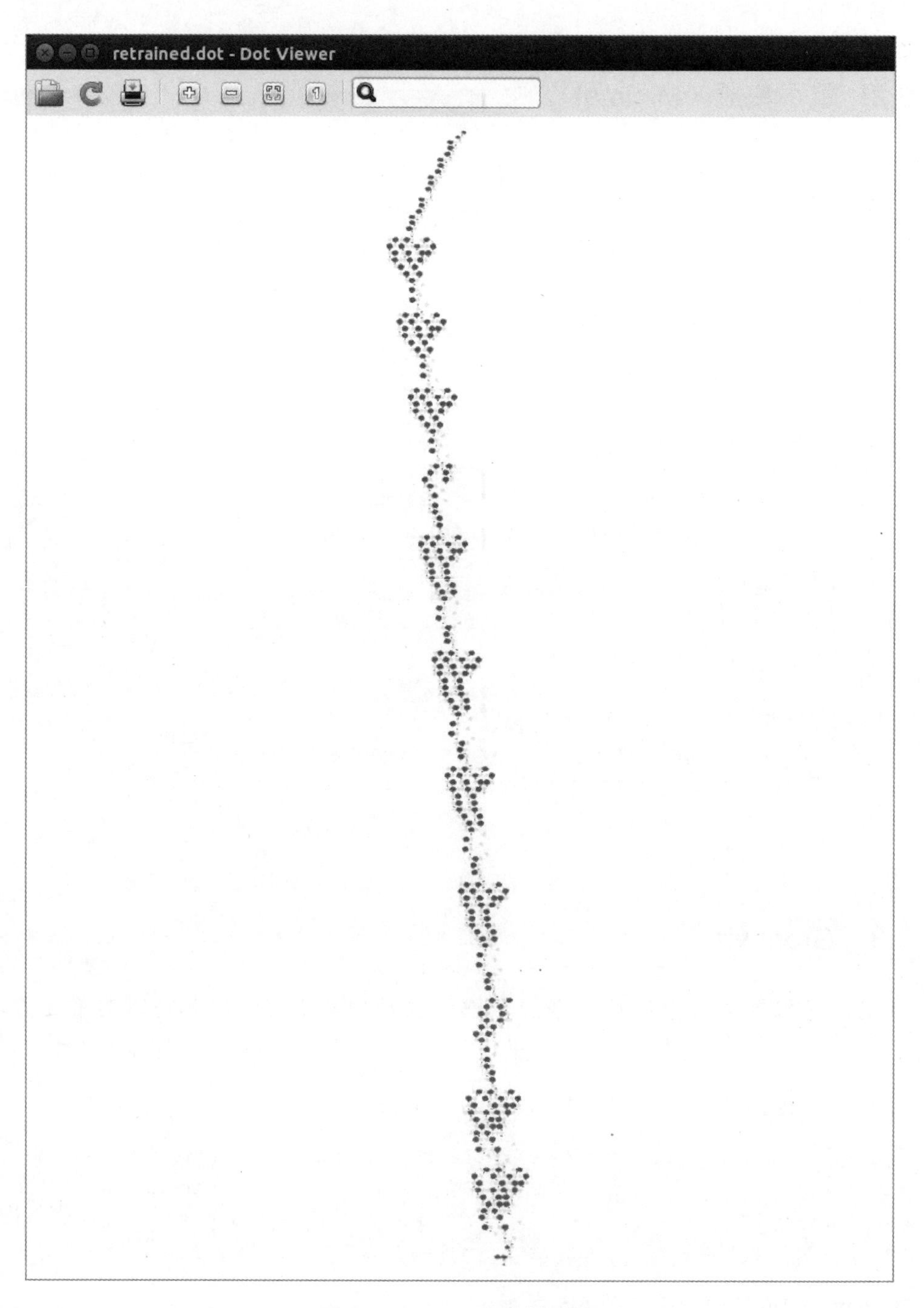

图 6-5　Graphviz 的模型图

模型的细节如图 6-6 所示。

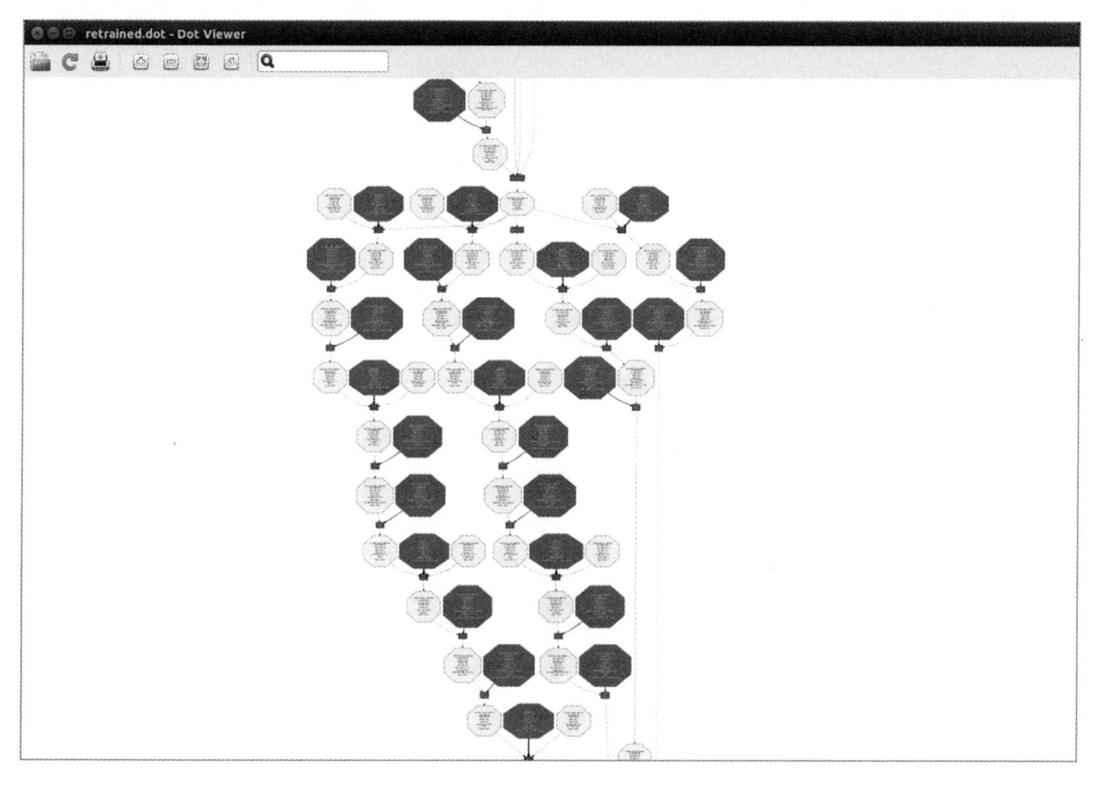

图 6-6　Graphviz 模型的细节图

接下来，让我们将 TensorFlow Lite 模型转换为 Graphviz：

```
    $ toco --input_file=tf_files/optimized_graph.lite --output_file=tf_files/
lite.dot  --input_format=TFLITE  --output_format=GRAPHVIZ_DOT  --input_shape=
1,224,224,3 --input_array=input --output_array=final_result --inference_type=
FLOAT --input_data_type=FLOAT
```

由于代码问题，程序不能正常执行，问题在这里：

```
    const auto& input_shape = input_array.shape();
    CHECK_EQ(input_shape.dimensions_count(), 4);
```

Toco 需在检查形状和形状尺寸之前检查缓冲区指针。tooling_util.cc 中 CheckEachArray（cnst Model& model）的基本逻辑是正确的，但缺少边界检查，解决方法是添加如下代码：

```
    if (!input_array.has_shape()) {
      return;
```

```
}
```

在此修复之后，在检查形状之前，重新运行下面的命令：

```
$ toco --input_file=tf_files/optimized_graph.lite --output_file=tf_files/
lite.dot  --input_format=TFLITE  --output_format=GRAPHVIZ_DOT  --input_shape=
1,224,224,3 --input_array=input --output_array=final_result --inference_type=
FLOAT --input_data_type=FLOAT
```

得到如图 6-7 所示的 TensorFlow Lite 的模型图。

图 6-7　TensorFlow Lite 的模型图

TensorFlow Lite 模型放大以后的细节如图 6-8 所示。

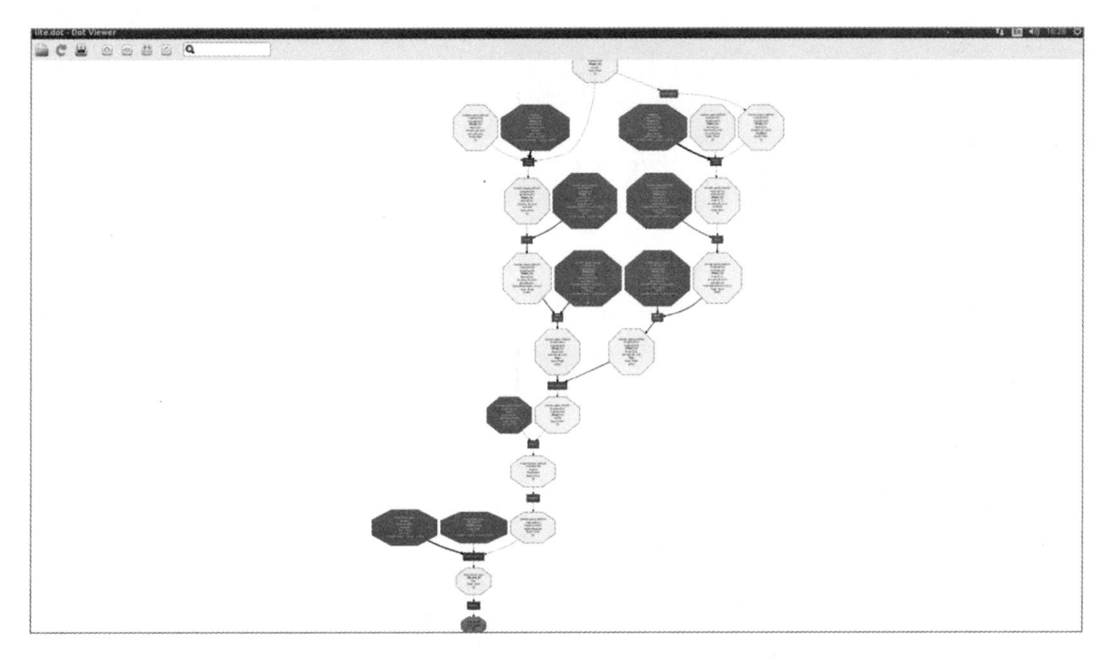

图 6-8　TensorFlow Lite 模型的细节图

下面我们把重新训练的模型生成可视图，代码如下：

```
    $ /opt/tensorflow/bin/toco/toco --input_file=tf_files/retrained_graph.pb
--output_file=tf_files/retrained.dot        --input_format=TENSORFLOW_GRAPHDEF
--output_format=GRAPHVIZ_DOT  --input_shape=1,224,224,3  --input_array=input
--output_array=final_result --inference_type=FLOAT --input_data_type=FLOAT
```

然后，在 Linux 上，使用 xdot 查看图表：

```
    xdot tf_files/retrained.dot
```

关于 xdot 的一个很酷的特性是，你可以单击一个块，这个块将被置中，然后你可以放大和缩小，也可以将 dot 文件转换为 pdf 文件。Toco 还提供了 dump_graphviz，可以在转换时自动生成 dot 文件，具体实现代码如下：

```
    /opt/tensorflow/bin/toco/toco    --input_file=tf_files/retrained_graph.pb
--output_file=/tmp/optimized_graph.lite    --input_format=TENSORFLOW_GRAPHDEF
--output_format=TFLITE      --input_shape=1,224,224,3       --input_array=input
--output_array=final_result  --inference_type=FLOAT  --input_data_type=FLOAT
--dump_graphviz=/tmp
```

代码执行完毕，生成 dot 文件如下：

```
-rw-rw-r-- 1 toco_AFTER_ALLOCATION.dot
-rw-rw-r-- 1 toco_AFTER_TRANSFORMATIONS.dot
-rw-rw-r-- 1 toco_AT_IMPORT.dot
```

对于生成的 dot 文件，读者可以用 xdot 命令进行查看。

6.3.4 模型评效

执行下面的命令，生成模型概要。

```
$ summarize_graph --in_graph=tf_files/retrained_graph.pb
```

输出结果如下：

```
Found 1 possible inputs: (name=Placeholder, type=float(1), shape=[?,299,299,3])
No variables spotted.
Found 1 possible outputs: (name=final_result, op=Softmax)
Found 21830264 (21.83M) const parameters, 0 (0) variable parameters, and 0 control_edges
Op types used: 490 Const, 378 Identity, 94 Conv2D, 94 FusedBatchNorm, 94 Relu, 15 ConcatV2, 9 AvgPool, 4 MaxPool, 1 Add, 1 Mean, 1 Mul, 1 Placeholder, 1 PlaceholderWithDefault, 1 MatMul, 1 Softmax, 1 Squeeze, 1 Sub
```

执行下面的命令，生成优化后的模型概要。

```
$ summarize_graph --in_graph=tensorflow_hub/tf_files/optimized_graph.pb
```

输出结果如下：

```
Found 1 possible inputs: (name=Placeholder, type=float(1), shape=None)
No variables spotted.
Found 1 possible outputs: (name=final_result, op=Softmax)
Found 21778616 (21.78M) const parameters, 0 (0) variable parameters, and 0 control_edges
Op types used: 208 Const, 94 BiasAdd, 94 Conv2D, 94 Relu, 15 ConcatV2, 9 AvgPool, 4 MaxPool, 1 Add, 1 MatMul, 1 Mean, 1 Mul, 1 Placeholder, 1 PlaceholderWithDefault, 1 Softmax, 1 Squeeze, 1 Sub
```

可以把模型与 tensorflow/tools/benchmark 一起使用，benchmark_model 的用例如下：

```
benchmark_model --graph=tf_files/retrained_graph.pb --show_flops --input_layer=Placeholder  --input_layer_type=float  --input_layer_shape=1,299,299,3 --output_layer=final_result
```

输出结果如下：

```
    2018-05-15 14:52:05.401180: I tensorflow/core/util/stat_summarizer.cc:468]
========================== Summary by node type ==============================
      2018-05-15 14:52:05.401186: I tensorflow/core/util/stat_summarizer.cc:
468]        [Node type]  [count]  [avg ms]    [avg %]     [cdf %]  [mem KB][times
called]
      2018-05-15 14:52:05.401195: I tensorflow/core/util/stat_summarizer.cc:
468]        Conv2D       94    69.622    76.439%    76.439% 35869.953       94
      2018-05-15 14:52:05.401203: I tensorflow/core/util/stat_summarizer.cc:
468]        AvgPool       9    8.866     9.734%    86.173% 8009.600        9
      2018-05-15 14:52:05.401211: I tensorflow/core/util/stat_summarizer.cc:
468]        BiasAdd      94    6.322     6.941%    93.114%    0.000       94
      2018-05-15 14:52:05.401220: I tensorflow/core/util/stat_summarizer.cc:
468]        MaxPool       4    2.669     2.930%    96.044% 2834.560        4
      2018-05-15 14:52:05.401227: I tensorflow/core/util/stat_summarizer.cc:
468]        Relu         63    1.813     1.991%    98.035%    0.000       63
      2018-05-15 14:52:05.401235: I tensorflow/core/util/stat_summarizer.cc:
468]        ConcatV2     15    1.086     1.192%    99.227% 10678.528       15
      2018-05-15 14:52:05.401243: I tensorflow/core/util/stat_summarizer.cc:
468]        Const       194    0.376     0.413%    99.640%    0.000      194
      2018-05-15 14:52:05.401251: I tensorflow/core/util/stat_summarizer.cc:
468]        Mul           1    0.126     0.138%    99.778% 1072.812        1
      2018-05-15 14:52:05.401259: I tensorflow/core/util/stat_summarizer.cc:
468]        Mean          1    0.057     0.063%    99.841%    8.192        1
      2018-05-15 14:52:05.401268: I tensorflow/core/util/stat_summarizer.cc:
468]        Sub           1    0.054     0.059%    99.900%    0.000        1
      2018-05-15 14:52:05.401276: I tensorflow/core/util/stat_summarizer.cc:
468]        NoOp          1    0.042     0.046%    99.946%    0.000        1
      2018-05-15 14:52:05.401284: I tensorflow/core/util/stat_summarizer.cc:
468]        MatMul        1    0.015     0.016%    99.963%    0.020        1
      2018-05-15 14:52:05.401292: I tensorflow/core/util/stat_summarizer.cc:
468]        _Retval       1    0.010     0.011%    99.974%    0.000        1
      2018-05-15 14:52:05.401300: I tensorflow/core/util/stat_summarizer.cc:
468]        Squeeze       1    0.006     0.007%    99.980%    0.000        1
      2018-05-15 14:52:05.401309: I tensorflow/core/util/stat_summarizer.cc:
468]        Softmax       1    0.006     0.007%    99.987%    0.000        1
      2018-05-15 14:52:05.401317: I tensorflow/core/util/stat_summarizer.cc:
468]        Add           1    0.006     0.007%    99.993%    0.000        1
      2018-05-15 14:52:05.401325: I tensorflow/core/util/stat_summarizer.cc:
468]        _Arg          1    0.004     0.004%    99.998%    0.000        1
```

```
    2018-05-15 14:52:05.401334: I tensorflow/core/util/stat_summarizer.cc:
468]  PlaceholderWithDefault      1    0.002    0.002%   100.000%    0.000   1
    2018-05-15 14:52:05.401342: I tensorflow/core/util/stat_summarizer.cc:
468]
    2018-05-15 14:52:05.401347: I tensorflow/core/util/stat_summarizer.cc:
468] Timings (microseconds): count=217 first=87776 curr=87457 min=84829
max=207830 avg=91316.4 std=18506
    2018-05-15 14:52:05.401353: I tensorflow/core/util/stat_summarizer.cc:
468] Memory (bytes): count=217 curr=58473664(all same)
    2018-05-15 14:52:05.401359: I tensorflow/core/util/stat_summarizer.cc:
468] 484 nodes observed
    2018-05-15 14:52:05.401364: I tensorflow/core/util/stat_summarizer.cc:
468]
    2018-05-15 14:52:06.708874: I tensorflow/tools/benchmark/benchmark_
model.cc:631] FLOPs estimate: 11.42B
    2018-05-15 14:52:06.708915: I tensorflow/tools/benchmark/benchmark_
model.cc:633] FLOPs/second: 260.55B
```

7

第 7 章 用 TensorFlow Lite 构建机器学习应用

本章主要介绍开发面向移动端机器学习应用的过程和方法。如果读者在移动应用程序中使用 TensorFlow Lite 模型，必须选择预先训练的模型或自定义的模型，将模型转换为 TensorFLow Lite 的模型格式，再将模型集成到应用程序中。

7.1 模型设计

根据使用场景，读者可以选择一种流行的开源模型，例如 InceptionV3 或 MobileNets，并使用自定义数据集重新训练这些模型，也可以构建自己的自定义模型。

7.1.1 使用预先训练的模型

这里我们先用视觉模型举例。MobileNets 是 TensorFlow 的以移动优先设计的计算机视觉模型系列中的一个，这个模型旨在有效地最大限度地提高准确性，同时考虑到设备或嵌入式应用程序的有限资源。

MobileNets 是小型、低延迟、低功耗模型，优化参数可以满足各种资源受限制的应用。这个模型可用于物体的分类、检测、嵌入和分割，类似于其他流行的大型模型，例如 Inception。

谷歌为 MobileNets 提供了 16 个经过预先培训的 ImageNet 物体分类检查点，可用于各种规模的移动应用项目。Inception-v3 是一种图像识别模型，可以实现相当高的准确度，能够识别 1000 个类别的物体对象，例如“斑马”和“洗碗机”等。该模型使用卷积神经网络从输入图像中提取一般特征，并基于具有完全连接和 Softmax 层的那些特征对它们进行分类。

“移动端智能回复（On Device Smart Reply）”也是一种面向移动设备的模型，通过建议与上下文相关的信息，为传入的文本消息提供一键式回复。该模型专为内存受限设备（如手表和手机）而构建，并已成功用于 Android Wear 上的智能回复。目前，此模型只支持 Android。

这些预先训练的模型都可以在 https://www.tensorflow.org/lite/models 上下载。

7.1.2 重新训练

TensorFlow 也提供了工具和脚本供开发者使用来进行模型的重新训练。比如，image_retraining 是一个图像再训练的工具。

首先，下载源代码：

```
$ git clone https://github.com/tensorflow/hub.git
```

然后，构建执行文件：

```
$ bazel build tensorflow/examples/image_retraining:retrain
```

TensorFlow Hub 似乎很久没有人更新和维护了，会出现编译错误。笔者个人的建议是，把需要的源代码复制到 TensorFlow 的仓里，并做适当的更改。比如，把下面的更改加进 WORKSPACE 里，就可以满足构建的依赖要求，具体代码如下：

```
load("@bazel_tools//tools/build_defs/repo:git.bzl", "git_repository")
git_repository(
```

```
        name = "protobuf_bzl",
        # v3.6.0
        commit = "ab8edf1dbe2237b4717869eaab11a2998541ad8d",
        remote = "https://github.com/google/protobuf.git",
    )
    bind(
        name = "protobuf",
        actual = "@protobuf_bzl//:protobuf",
    )
    bind(
        name = "protobuf_python",
        actual = "@protobuf_bzl//:protobuf_python",
    +)
    bind(
        name = "protobuf_python_genproto",
        actual = "@protobuf_bzl//:protobuf_python_genproto",
    )
    bind(
        name = "protoc",
        actual = "@protobuf_bzl//:protoc",
    )
    # Using protobuf version 3.6.0
    http_archive(
        name = "com_google_protobuf",
        strip_prefix = "protobuf-3.6.0",
        urls = ["https://github.com/google/protobuf/archive/v3.6.0.zip"],
    )
```

上述代码编译后，就得到可执行文件。在 retrain.py 里可以看到如下所示的代码，功能是把模型的参数传给执行文件。

```
    parser.add_argument(
        '--tfhub_module',
        type=str,
        default=(
            'https://tfhub.dev/google/imagenet/inception_v3/feature_
vector/1'),
            help="""
            Which TensorFlow Hub module to use.
            See    https://github.com/tensorflow/hub/blob/master/docs/modules/
image.md
            for some publicly available ones.
```

```
        """)
```

准备好数据文件后，就可以进行图像的再训练。首先，把图片存放到一个文件夹（这里我们把图像存储到 flower_photos）中。然后，执行脚本，它的中间输出将在系统 tmp 文件夹中的 tfhub_modules 中。下面的脚本分别对三个模型进行再训练，这三个模型分别是 inception_v3、mobilenet 和 mobilenet 的定点数模型。

```
    $ bazel-bin/examples/image_retraining/retrain --image_dir flower_photos
--tfhub_module https://tfhub.dev/google/imagenet/inception_v3/feature_vector/1
--saved_model_dir inception_v3/saved_model/ --output_graph inception_v3/
retain_graph.pb --output_labels inception_v3/label.txt --summaries_dir
inception_v3/summaries/ --bottleneck_dir inception_v3/bottleneck/

    $ bazel-bin/examples/image_retraining/retrain --image_dir flower_photos
--tfhub_module https://tfhub.dev/google/imagenet/mobilenet_v1_100_224/feature_
vector/1 --saved_model_dir mobilenet_float/saved_model/ --output_graph
mobilenet_float/retain_graph.pb --output_labels mobilenet_float/label.txt
--summaries_dir mobilenet_float/summaries/ --bottleneck_dir mobilenet_float/
bottleneck/

    $ bazel-bin/examples/image_retraining/retrain --image_dir flower_photos
--tfhub_module https://tfhub.dev/google/imagenet/mobilenet_v1_100_224/
quantops/feature_vector/1 --saved_model_dir mobilenet_quant/saved_model/
--output_graph mobilenet_quant/retain_graph.pb --output_labels mobilenet_
quant/label.txt --summaries_dir mobilenet_quant/summaries/ --bottleneck_dir
mobilenet_quant/bottleneck/
```

7.1.3 使用瓶颈（Bottleneck）

根据机器速度的不同，上面脚本可能需要很长时间才能完成训练。执行的第一个阶段是分析存储的所有图像，并计算和缓存每个映像的瓶颈值。“瓶颈”是一个非正式术语，实践中我们经常使用它，它表示在最终输出层之前的层（在 TensorFlow Hub 里将其称为“图像特征向量”）。此倒数第二层经过训练，已经可以满足区分的要求，进而输出一组足够好的结果。这意味着它必须包含一个有意义且紧凑的图像概要，即它必须包含足够的信息，以便分类器在一组非常有限的值中做出正确的选择。我们把最后一层经过再训练就可以用于新类的原因就在于此。结果表明，用于区分 ImageNet 中所有 1000 个类所需的信息通常也可用于区分新的类型。

因为每个图像在训练期间多次重复使用，并且计算每个瓶颈耗时很长，如果我们能把瓶颈的值缓存到磁盘上，就可以节省大量时间。默认情况下，它们存储在/tmp/bottleneck目录中，如果重新运行脚本，它们可以被重用。使用瓶颈进行训练的实现代码如下：

```
    IMAGE_SIZE=224
    ARCHITECTURE="mobilenet_0.50_${IMAGE_SIZE}"
    $ bazel-bin/examples/image_retraining/retrain --bottleneck_dir=tf_files/
bottlenecks    --how_many_training_steps=500    --model_dir=tf_files/models/
--summaries_dir=tf_files/training_summaries/"${ARCHITECTURE}"
--output_graph=tf_files/retrained_graph.pb
--output_labels=tf_files/retrained_labels.txt
--architecture="${ARCHITECTURE}" --image_dir=flower_photos
```

我们现在可以使用训练好的模型，尝试去检测一张图像，具体实现代码如下：

```
    $ label_image --graph=tf_files/retrained_graph.pb --image=flower_photos/
sunflowers/24459548_27a783feda.jpg --input_layer=Placeholder --output_layer=
final_result    --labels=tf_files/retrained_labels.txt    --input_width=299
--input_height=299
```

运行结果如下：

```
    2018-05-10   15:53:16.354204:   I   tensorflow/core/platform/cpu_feature_
guard.cc:141] Your CPU supports instructions that this TensorFlow binary was not
compiled to use: SSE4.1 SSE4.2 AVX AVX2 FMA
    2018-05-10 15:53:17.986350: I tensorflow/examples/label_image/main.cc:251]
sunflowers (3): 0.825468
    2018-05-10 15:53:17.986390: I tensorflow/examples/label_image/main.cc:251]
tulips (4): 0.0628504
    2018-05-10 15:53:17.986401: I tensorflow/examples/label_image/main.cc:251]
daisy (0): 0.0562632
    2018-05-10 15:53:17.986408: I tensorflow/examples/label_image/main.cc:251]
dandelion (1): 0.0406035
    2018-05-10 15:53:17.986414: I tensorflow/examples/label_image/main.cc:251]
roses (2): 0.0148145
```

现在，读者还可以使用优化器，对生成的模型进行进一步的优化：

```
    $ bazel build tensorflow/python/tools:optimize_for_inference
    $  bazel-bin/tensorflow/python/tools/optimize_for_inference  --input=tf_
files/retrained_graph.pb --output=tf_files/optimized_graph.pb --input_name=
"Placeholder" --output_name="final_result"
```

优化的效果如下：

```
    2018-05-10 16:16:26.601517: I tensorflow/lite/toco/graph_transformations/
```

```
graph_transformations.cc:39] Before Removing unused ops: 1072 operators, 1658 arrays (0 quantized)
    2018-05-10 16:16:26.631377: I tensorflow/lite/toco/graph_transformations/graph_transformations.cc:39] Before general graph transformations: 1072 operators, 1658 arrays (0 quantized)
    2018-05-10 16:16:26.743531: I tensorflow/lite/toco/graph_transformations/graph_transformations.cc:39] After general graph transformations pass 1: 128 operators, 323 arrays (0 quantized)
    2018-05-10 16:16:26.745757: I tensorflow/lite/toco/graph_transformations/graph_transformations.cc:39] After general graph transformations pass 2: 126 operators, 319 arrays (0 quantized)
    2018-05-10 16:16:26.747896: I tensorflow/lite/toco/graph_transformations/graph_transformations.cc:39] Before dequantization graph transformations: 126 operators, 319 arrays (0 quantized)
    2018-05-10 16:16:26.749797: I tensorflow/lite/toco/allocate_transient_arrays.cc:329] Total transient array allocated size: 0 bytes, theoretical optimal value: 0 bytes.
```

现在，我们可以检验一下训练过的模型，进而确定效果。这里我们使用 label_image 工具：

```
    $ label_image --graph=tf_files/optimized_graph.pb --image=flower_photos/sunflowers/24459548_27a783feda.jpg --input_layer=Placeholder --output_layer=final_result --labels=tf_files/retrained_labels.txt --input_width=299 --input_height=299
```

输出结果如下：

```
    2018-05-10 16:11:33.589364: I tensorflow/core/platform/cpu_feature_guard.cc:141] Your CPU supports instructions that this TensorFlow binary was not compiled to use: SSE4.1 SSE4.2 AVX AVX2 FMA
    2018-05-10 16:11:35.120186: I tensorflow/examples/label_image/main.cc:251] sunflowers (3): 0.825468
    2018-05-10 16:11:35.120239: I tensorflow/examples/label_image/main.cc:251] tulips (4): 0.0628507
    2018-05-10 16:11:35.120263: I tensorflow/examples/label_image/main.cc:251] daisy (0): 0.0562634
    2018-05-10 16:11:35.120269: I tensorflow/examples/label_image/main.cc:251] dandelion (1): 0.0406035
    2018-05-10 16:11:35.120276: I tensorflow/examples/label_image/main.cc:251] roses (2): 0.0148145
```

确认结果之后，就可以使用 Toco 工具把训练好的模型转换成 TensorFlow Lite 的模型。如果你的预装 Toco 有以下问题，你可以重新编译一下。

```
$ toco
TOCO from pip install is currently not working on command line.
Please use the python TOCO API or use
bazel run tensorflow/lite:toco -- <args> from a TensorFlow source dir.
```

执行下面的命令，就可以重新编译 Toco。

```
$ bazel build tensorflow/lite/toco:toco
```

然后我们就可以使用 Toco 了。执行下面的命令，把 inception_v3 和 mobilenet 的模型转换成 TensorFlow Lite 的模型。

```
$ /opt/tensorflow/bin/toco/toco --input_file=inception_v3/retain_graph.pb
--output_file=inception_v3/optimized_graph.lite
--input_format=TENSORFLOW_GRAPHDEF --output_format=TFLITE --input_shape=1,224,
224,3 --input_array=input --output_array=final_result --inference_type=FLOAT
--input_data_type=FLOAT --dump_graphviz=inception_v3/

$   /opt/tensorflow/bin/toco/toco   --input_file=mobilenet_float/retain_
graph.pb  --output_file=mobilenet_float/optimized_graph.lite  --input_format=
TENSORFLOW_GRAPHDEF --output_format=TFLITE --input_shape=1,224,224,3 --input_
array=input --output_array=final_result --inference_type=FLOAT --input_data_
type=FLOAT --dump_graphviz=mobilenet_float/
```

上面我们把预先训练的模型在自己的数据集上进行训练，原始的模型可以对 1000 个类进行分类。如果这些类不足以满足你的使用需求，则需要重新训练模型。这种技术称为迁移学习，从已经训练过问题的模型开始，在类似的问题上重新训练模型。从头开始深度学习可能需要数天时间，但迁移学习相当快。为了进行迁移学习，读者需要生成标有相关类的自定义数据集。

开发人员也可以选择使用 TensorFlow 先训练自定义模型。即先编写模型，再对数据进行训练。

使用 TensorFlow Lite 在移动端上的开发和使用 TensorFlow Mobile 非常类似。一般来讲，应用程序的代码重用度非常高，要做的事基本就是，根据模型的不同对模型的输入和输出做出调整。

前面，我们对如何构建应用做了很多讲解，这里就不重复了。TensorFlow Lite 支持 TensorFlow 运算符的子集。有关支持的运算符及其用法，可以参阅 TensorFlow 的官方文档。TensorFlow Lite 也认识到这个问题，谷歌正在进行开发，同时也在研究可以快速使用已有算子的方法。

7.2 开发应用

这里我们介绍开发 TensorFlow Lite 应用所使用的接口和一些技术要点。

7.2.1 程序接口

TensorFlow Lite 为开发人员提供了非常容易使用的 API 接口，只需几步就可以完成一个简单应用的开发。下面是示例代码：

```
// 定义 TfLite
private Interpreter tfLite;

// Tflite，创建实例
try {
  tfLite = new Interpreter(loadModelFile(MODEL_FILE));
} catch (IOException e) {
  Log.e(TAG, "Failed to create TFLite");
  finish();
}

// 运行 Tflite

tfLite.run(data, label);
```

只需简单的三步就可以运行应用。第一步，定义一个解释器；第二步，生成一个实例；第三步，运行。

TensorFlow Lite 提供了 6 个构建函数可以生成一个解释器。比如，应用可以直接传入一个文件的对象：

```
/**
 * Initializes a {@code Interpreter}
 *
 * @param modelFile: a File of a pre-trained TF Lite model.
 */
public Interpreter(@NonNull File modelFile) {
  this(modelFile, /*options = */ null);
}
```

笔者推荐读者使用下面的 API 去生成解释器：

```
    /**
     * Initializes a {@code Interpreter} with a {@code ByteBuffer} of a model
file and a set of custom
     * {@link #Options}.
     *
     * <p>The ByteBuffer should not be modified after the construction of a {@code
Interpreter}. The
     * {@code ByteBuffer} can be either a {@code MappedByteBuffer} that
memory-maps a model file, or a
     * direct {@code ByteBuffer} of nativeOrder() that contains the bytes content
of a model.
     */
    public Interpreter(@NonNull ByteBuffer byteBuffer, Options options) {
      wrapper = new NativeInterpreterWrapper(byteBuffer, options);
    }
```

这种生成解释器的方法的优点是，可以减少模型加载时间。建议使用内存共享，先生成 MappedByteBuffer，再传入解析器。

对于运行函数，有两个选择，即 run 接口和 runForMultipleInputsOutputs 接口。这两个接口区别不大。如果模型只有一个输入，就用 run 接口；如果模型有多个输入，就用 runForMultipleInputsOutputs 接口。

接口 run 的代码如下：

```
    /**
     * Runs model inference if the model takes only one input, and provides only
one output.
     *
     * <p>Warning: The API runs much faster if {@link ByteBuffer} is used as input
data type. Please
     * consider using {@link ByteBuffer} to feed primitive input data for better
performance.
     *
     * @param input an array or multidimensional array, or a {@link ByteBuffer}
of primitive types
     *     including int, float, long, and byte. {@link ByteBuffer} is the
preferred way to pass large
     *     input data for primitive types, whereas string types require using the
(multi-dimensional)
     *     array input path. When {@link ByteBuffer} is used, its content should
```

```
remain unchanged
     *     until model inference is done. A {@code null} value is allowed only
if the caller is using
     *     a {@link Delegate} that allows buffer handle interop, and such a buffer
has been bound to
     *     the input {@link Tensor}.
     * @param output a multidimensional array of output data, or a {@link
ByteBuffer} of primitive
     *     types including int, float, long, and byte. A null value is allowed
only if the caller is
     *     using a {@link Delegate} that allows buffer handle interop, and such
a buffer has been
     *         bound to the output {@link Tensor}. See also {@link
Options#setAllowBufferHandleOutput()}.
     */
    public void run(Object input, Object output) {
      Object[] inputs = {input};
      Map<Integer, Object> outputs = new HashMap<>();
      outputs.put(0, output);
      runForMultipleInputsOutputs(inputs, outputs);
    }
```

接口 runForMultipleInputsOutputs 的代码如下：

```
    /**
    /**
     * Runs model inference if the model takes multiple inputs, or returns multiple
outputs.
     *
     * <p>Warning: The API runs much faster if {@link ByteBuffer} is used as input
data type. Please
     * consider using {@link ByteBuffer} to feed primitive input data for better
performance.
     *
     * <p>Note: {@code null} values for invididual elements of {@code inputs}
and {@code outputs} is
     * allowed only if the caller is using a {@link Delegate} that allows buffer
handle interop, and
     * such a buffer has been bound to the corresponding input or output {@link
Tensor}(s).
     *
     * @param inputs an array of input data. The inputs should be in the same
order as inputs of the
     *     model. Each input can be an array or multidimensional array, or a {@link
```

```
ByteBuffer} of
     *     primitive types including int, float, long, and byte. {@link ByteBuffer}
is the preferred
     *     way to pass large input data, whereas string types require using the
(multi-dimensional)
     *     array input path. When {@link ByteBuffer} is used, its content should
remain unchanged
     *     until model inference is done.
     * @param outputs a map mapping output indices to multidimensional arrays
of output data or {@link
     *     ByteBuffer}s of primitive types including int, float, long, and byte.
It only needs to keep
     *     entries for the outputs to be used.
     */
    public void runForMultipleInputsOutputs(
       @NonNull Object[] inputs, @NonNull Map<Integer, Object> outputs) {
      checkNotClosed();
      wrapper.run(inputs, outputs);
    }
```

下面是一个输入运行函数对应的原生 C 的实现：

```
    public void run(@NonNull Object input, @NonNull Object output) {
      Object[] inputs = {input};
      Map<Integer, Object> outputs = new HashMap<>();
      outputs.put(0, output);
      runForMultipleInputsOutputs(inputs, outputs);
    }
```

下面是多个输入运行函数对应的原生 C 代码的实现，读者可以参考。

```
    public void runForMultipleInputsOutputs(
      @NonNull Object[] inputs, @NonNull Map<Integer, Object> outputs) {
      if (wrapper == null) {
        throw new IllegalStateException("Internal error: The Interpreter has
already been closed.");
      }
      Tensor[] tensors = wrapper.run(inputs);
      if (outputs == null || tensors == null || outputs.size() > tensors.length) {
        throw new IllegalArgumentException("Output error: Outputs do not match
with model outputs.");
      }
      final int size = tensors.length;
      for (Integer idx : outputs.keySet()) {
      if (idx == null || idx < 0 || idx >= size) {
```

```
        throw new IllegalArgumentException(
        String.format(
            "Output error: Invalid index of output %d (should be in range [0, %d))",
            idx, size));
        }
        tensors[idx].copyTo(outputs.get(idx));
      }
    }
```

7.2.2 线程和性能

TensorFlow Lite 是单线程的运行程序。它会根据模型按顺序执行。但是，在 Interpreter.java 里，TensorFlow Lite 提供了设定线程的接口，现在来看一下它的内部实现：

```
interpreter->SetNumThreads(static_cast<int>(num_threads));
```

在 C++的实现过程中，程序把线程数传给了 eigen。现在，我们看到了，线程数是针对 eigen 的，不是针对 TensorFlow Lite 的。具体代码如下：

```
void Interpreter::SetNumThreads(int num_threads) {
  context_.recommended_num_threads = num_threads;

  gemm_support::SetNumThreads(&context_, num_threads);
  eigen_support::SetNumThreads(&context_, num_threads);
}
```

下面的代码是解释器执行过程的实现，可以看到这是一个单线程的执行：

```
for (int execution_plan_index = 0;
    execution_plan_index < execution_plan_.size(); execution_plan_index++) {
  if (execution_plan_index == next_execution_plan_index_to_prepare_) {
    TF_LITE_ENSURE_STATUS(PrepareOpsAndTensors());
    TF_LITE_ENSURE(&context_, next_execution_plan_index_to_prepare_ >=
        execution_plan_index);
    }
    int node_index = execution_plan_[execution_plan_index];
    TfLiteNode& node = nodes_and_registration_[node_index].first;
    const TfLiteRegistration& registration =
        nodes_and_registration_[node_index].second;
    **  __android_log_print(ANDROID_LOG_VERBOSE,  "TFLITE",  "node:%d",
registration.builtin_code);**
    SCOPED_OPERATOR_PROFILE(profiler_, node_index);
```

现在，来看一下多线程对运行的影响，我们可以使用开源的图像检测器做测试，分别使用 1 个线程和 10 个线程，测试结果如下。

1 个线程：

```
App                              CPU float        CPU quant        NPU
image classifier / frame           800ms            150ms          70ms
```

10 个线程：

```
App                              CPU float       CPU quant         NPU
image classifier / frame           800ms            150ms          70ms
```

看上去好像差别并不大。从其他测试来看，对有些应用的性能提升不是很大。注意，这里的多线程主要是软件的实现。

7.2.3　模型优化

在移动设备上运行机器学习应用的一个关键是优化模型。有很多方法可以优化模型，比如定点化、模型剪裁、并行化处理等。在 TensorFlow 出来之前，就已经有了为移动设备等小型运算设备优化的模型，我们在学习 TensorFlow 怎样为移动设备设计之外，还要检查一下这些模型的内部结构，看看它们的实际效果，并为后面的模型优化做准备。

1. 量子化模型

我们先来看一下 Mobile Net，这是一个为移动设备等做优化的模型。这次，我们不用直接下载的方法，而直接从 Android 的应用中检验模型。通过这个例子，我也希望读者可以思考怎样保护自己的模型。

首先，编译 TensorFlow Lite 的应用：

```
    $ bazel build -c opt --cxxopt=--std=c++11 --fat_apk_cpu=arm64-v8a
//tensorflow/lite/java/demo/app/src/main:TfLiteCameraDemo
```

假设我们从网上下载了这个应用，那么我们可以先解压 APK：

```
$ unzip TfLiteCameraDemo.apk
```

然后，通过检查应用的文件，找到模型文件：

```
$ ls -l assets/
total 4220
-rw-rw-rw- 1  BUILD
-rw-rw-rw- 1  labels_imagenet_slim.txt
```

```
    -rw-rw-rw- 1  labels_mobilenet_quant_v1_224.txt
    -rw-rw-rw- 1  labels.txt
    -rw-rw-rw- 1  mobilenet_quant_v1_224.tflite
    -rw-rw-rw- 1  WORKSPACE
```

接着，使用 Toco 工具将其转换为 Graphviz 文件：

```
    $ toco --input_file=mobilenet_quant_v1_224.tflite --output_file=mobilenet_
quant_v1_224.dot --input_format=TFLITE  --output_format=GRAPHVIZ_DOT
```

运行上面的命令后得到了如图 7-1 所示的 Mobile Net 模型图。

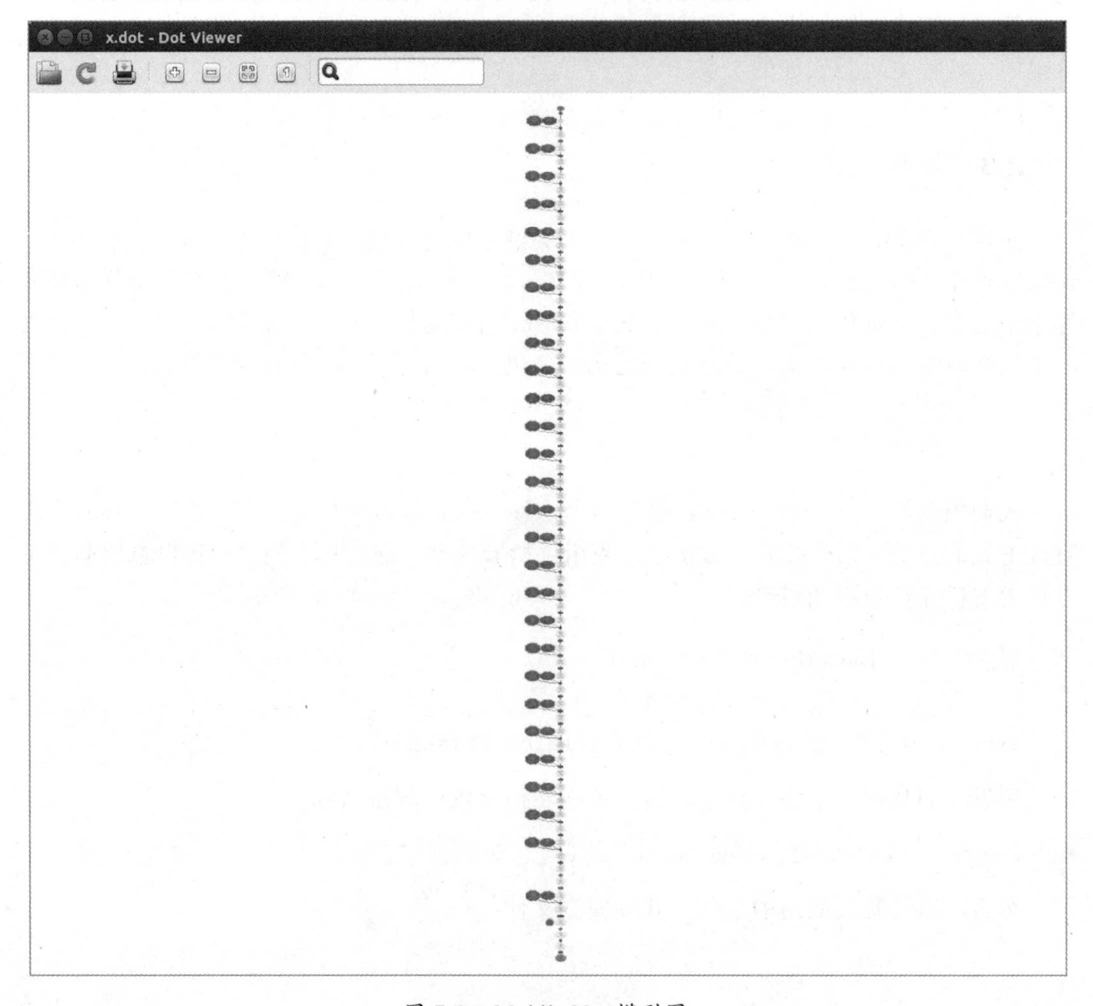

图 7-1　Mobile Net 模型图

最后，在 Camera2BasicFragment.java 中按照上面的示例代码启用调用日志记录，重新构建并运行它，我们应该在 Android logcat 中看到有类似的内容：

```
invoke start
node:3 kTfLiteBuiltinConv2d
node:4 kTfLiteBuiltinDepthwiseConv2d
node:3 kTfLiteBuiltinConv2d
node:4 kTfLiteBuiltinDepthwiseConv2d
node:3 kTfLiteBuiltinConv2d
node:4 kTfLiteBuiltinDepthwiseConv2d
node:3 kTfLiteBuiltinConv2d
node:4 kTfLiteBuiltinDepthwiseConv2d
node:3 kTfLiteBuiltinConv2d
node:4 kTfLiteBuiltinDepthwiseConv2d
node:3 kTfLiteBuiltinConv2d
node:4 kTfLiteBuiltinDepthwiseConv2d
node:3 kTfLiteBuiltinConv2d
node:4 kTfLiteBuiltinDepthwiseConv2d
node:3 kTfLiteBuiltinConv2d
node:4 kTfLiteBuiltinDepthwiseConv2d
node:3 kTfLiteBuiltinConv2d
node:4 kTfLiteBuiltinDepthwiseConv2d
node:3 kTfLiteBuiltinConv2d
node:4 kTfLiteBuiltinDepthwiseConv2d
node:3 kTfLiteBuiltinConv2d
node:4 kTfLiteBuiltinDepthwiseConv2d
node:3 kTfLiteBuiltinConv2d
node:4 kTfLiteBuiltinDepthwiseConv2d
node:3 kTfLiteBuiltinConv2d
node:4 kTfLiteBuiltinDepthwiseConv2d
node:3 kTfLiteBuiltinConv2d
node:1 kTfLiteBuiltinAveragePool2d
node:3 kTfLiteBuiltinConv2d
node:22 kTfLiteBuiltinReshape
node:25 kTfLiteBuiltinSoftmax
invoke end
```

一共 31 个 Ops，这些 Ops 是在 builtin_ops.h 中定义的，现在我们可以做交叉参考，可以用 logcat 直接输出 Ops 的名称。在上面的输出中，基本都是 kTfLiteBuiltinConv2d 和 kTfLiteBuiltinDepthwiseConv2d。从结构图中我们也可以看到这个模型是非常简单和高效的。

2. 浮点数模型

下面让我们来看一下 Inception v3 模型的内在结构，读者可以使用上面的方法，也可以从网站下载：

https://storage.googleapis.com/download.tensorflow.org/models/tflite/inception_v3_slim_2016_android_2017_11_10.zip

下载后解压缩，得到以下文件：

```
inflating: inceptionv3_slim_2016.tflite
inflating: imagenet_slim_labels.txt
```

我们使用 Toco，用同样的方法，可以得到模型的视图，代码如下：

```
toco --input_file=tensorflow/lite/java/demo/app/src/main/assets/
inceptionv3_slim_2016.tflite --output_file=lite.dot --input_format=TFLITE
--output_format=GRAPHVIZ_DOT
```

Inceptionv3 模型如图 7-2 所示。

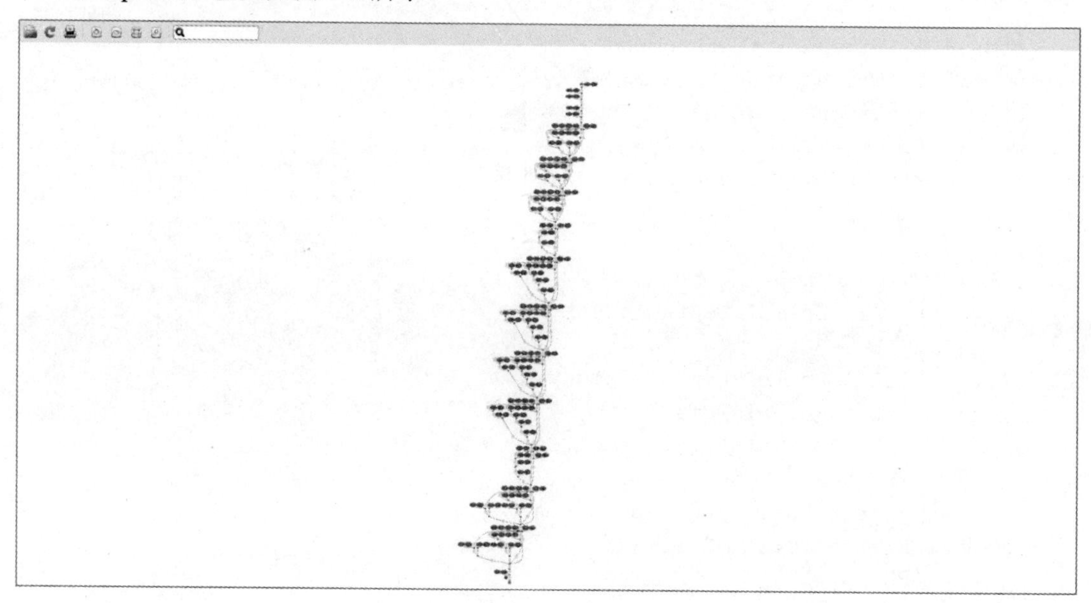

图 7-2　Inceptionv3 模型图

这个模型图（图 7-2）比 MobileNet 复杂得多。让我们放大这个图的一部分，看看解释器如何执行图。在图 7-2 中，单个输入是 MaxPool 的输出，单个输出是激活。需要连接

4 个输入，一共包括 6 个 conv2d。从图 7-2 中，我们可以很容易地发现 6 个 conv2d 操作可以并行执行。但是，从后面的 logcat 可以看出，interperter 是顺序执行它们的。模型放大后的细节如图 7-3 所示。

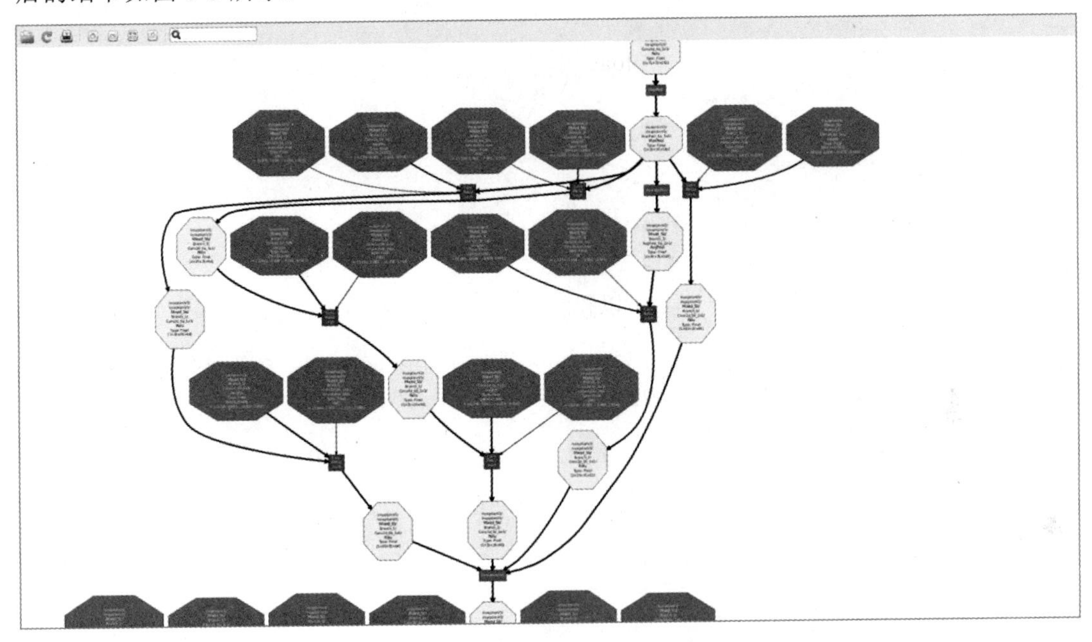

图 7-3　Inceptionv3 模型放大图

同样，按照上面的方法启用及调用日志记录，并运行它，我们应该在 logcat 中有类似的内容：

```
invoke start
node:3  kTfLiteBuiltinConv2d
node:3  kTfLiteBuiltinConv2d
node:3  kTfLiteBuiltinConv2d
node:17 kTfLiteBuiltinMaxPool2d
node:3  kTfLiteBuiltinConv2d
node:17 kTfLiteBuiltinMaxPool2d
node:3  kTfLiteBuiltinConv2d
node:3  kTfLiteBuiltinConv2d
node:3  kTfLiteBuiltinConv2d
node:3  kTfLiteBuiltinConv2d
node:3  kTfLiteBuiltinConv2d
node:1  kTfLiteBuiltinAveragePool2d
node:3  kTfLiteBuiltinConv2d
```

```
node:2  kTfLiteBuiltinConcatenation
node:3  kTfLiteBuiltinConv2d
node:3  kTfLiteBuiltinConv2d
node:1  kTfLiteBuiltinAveragePool2d
node:3  kTfLiteBuiltinConv2d
node:2  kTfLiteBuiltinConcatenation
node:3  kTfLiteBuiltinConv2d
node:3  kTfLiteBuiltinConv2d
node:3  kTfLiteBuiltinConv2d
node:3  kTfLiteBuiltinConv2d
node:1  kTfLiteBuiltinAveragePool2d
node:3  kTfLiteBuiltinConv2d
node:2  kTfLiteBuiltinConcatenation
node:3  kTfLiteBuiltinConv2d
node:3  kTfLiteBuiltinConv2d
node:3  kTfLiteBuiltinConv2d
node:3  kTfLiteBuiltinConv2d
node:17 kTfLiteBuiltinMaxPool2d
node:2  kTfLiteBuiltinConcatenation
node:3  kTfLiteBuiltinConv2d
node:3  kTfLiteBuiltinConv2d
node:3  kTfLiteBuiltinConv2d
node:3  kTfLiteBuiltinConv2d
node:3  kTfLiteBuiltinConv2d
node:3  kTfLiteBuiltinConv2d
node:1  kTfLiteBuiltinAveragePool2d
node:3  kTfLiteBuiltinConv2d
node:2  kTfLiteBuiltinConcatenation
node:3  kTfLiteBuiltinConv2d
node:3  kTfLiteBuiltinConv2d
node:3  kTfLiteBuiltinConv2d
node:3  kTfLiteBuiltinConv2d
node:3  kTfLiteBuiltinConv2d
node:3  kTfLiteBuiltinConv2d
node:3  kTfLiteBuiltinConv2d
node:1  kTfLiteBuiltinAveragePool2d
node:3  kTfLiteBuiltinConv2d
node:2  kTfLiteBuiltinConcatenation
node:3  kTfLiteBuiltinConv2d
node:3  kTfLiteBuiltinConv2d
node:1  kTfLiteBuiltinAveragePool2d
node:3  kTfLiteBuiltinConv2d
```

```
node:2  kTfLiteBuiltinConcatenation
node:3  kTfLiteBuiltinConv2d
node:3  kTfLiteBuiltinConv2d
node:3  kTfLiteBuiltinConv2d
node:3  kTfLiteBuiltinConv2d
node:3  kTfLiteBuiltinConv2d
node:1  kTfLiteBuiltinAveragePool2d
node:3  kTfLiteBuiltinConv2d
node:2  kTfLiteBuiltinConcatenation
node:3  kTfLiteBuiltinConv2d
node:3  kTfLiteBuiltinConv2d
node:17 kTfLiteBuiltinMaxPool2d
node:2  kTfLiteBuiltinConcatenation
node:3  kTfLiteBuiltinConv2d
node:3  kTfLiteBuiltinConv2d
node:2  kTfLiteBuiltinConcatenation
node:3  kTfLiteBuiltinConv2d
node:3  kTfLiteBuiltinConv2d
node:2  kTfLiteBuiltinConcatenation
node:1  kTfLiteBuiltinAveragePool2d
node:3  kTfLiteBuiltinConv2d
node:2  kTfLiteBuiltinConcatenation
node:3  kTfLiteBuiltinConv2d
node:3  kTfLiteBuiltinConv2d
node:2  kTfLiteBuiltinConcatenation
node:3  kTfLiteBuiltinConv2d
node:3  kTfLiteBuiltinConv2d
node:3  kTfLiteBuiltinConv2d
node:3  kTfLiteBuiltinConv2d
node:2  kTfLiteBuiltinConcatenation
node:1  kTfLiteBuiltinAveragePool2d
node:3  kTfLiteBuiltinConv2d
node:2  kTfLiteBuiltinConcatenation
node:1  kTfLiteBuiltinAveragePool2d
node:3  kTfLiteBuiltinConv2d
node:22 kTfLiteBuiltinReshape
```

可以看到一共有 61 个运算，所以这个模型比 MobileNet 复杂得多。

另外一种方法，我们还可以使用 Android Trace 来可视化运行时的信息。在 Interpreter.java 中，我们进行了以下修改：

```
public void run(@NonNull Object input, @NonNull Object output) {
```

```
    Trace.beginSection("run");
    Object[] inputs = {input};
    Map<Integer, Object> outputs = new HashMap<>();
    outputs.put(0, output);
    runForMultipleInputsOutputs(inputs, outputs);
    Trace.endSection();
  }
```

运行应用后，可以得到 Trace 文件。再使用下面的命令，把结果转换成网页。

```
python /opt/Android/sdk/platform-tools/systrace/systrace.py --app com.
example.android.tflitecamerademo -t 5 -o result.html
```

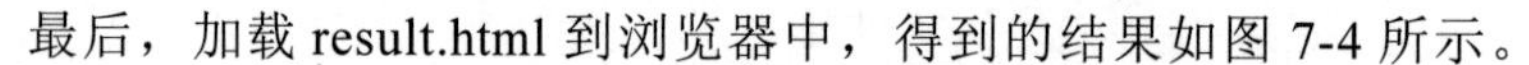

最后，加载 result.html 到浏览器中，得到的结果如图 7-4 所示。

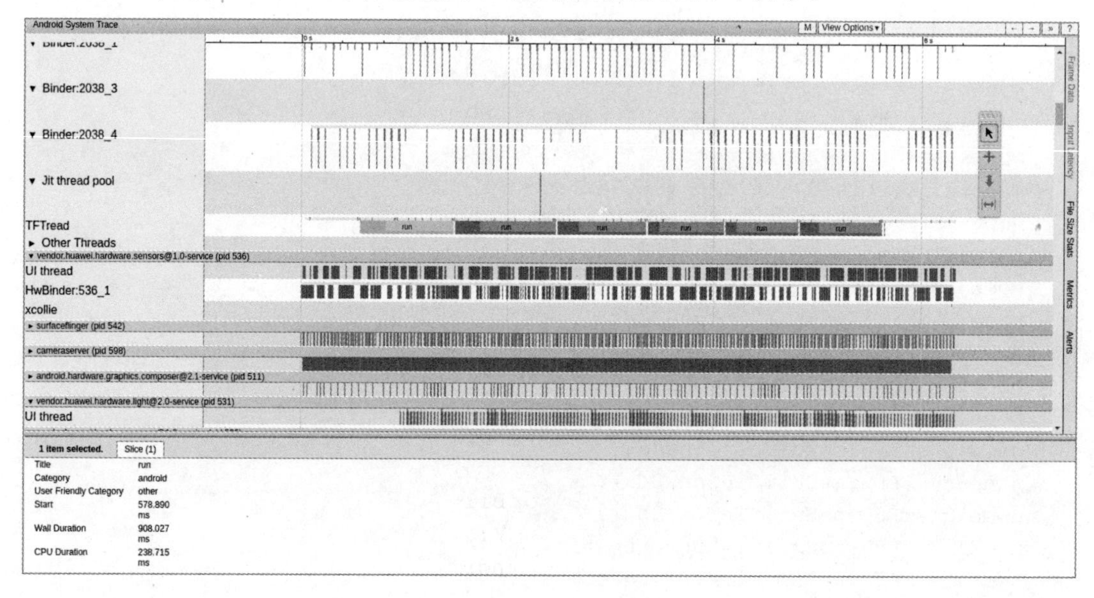

图 7-4　Trace 图

从图 7-4 中读者可以进一步分析运行的效能和优化点。

7.3 TensorFlow Lite 的应用

TensorFlow Lite 提供了几个应用供读者参考，这几个应用也可以作为开发人员开发产品的基础。有意思的是，TensorFlow Lite 的应用例子存放在两个文件夹中，一个文件夹是 tensorflow/lite/java/下的 demo，另一个是 tensorflow/lite/examples/下的 android。

tensorflow/lite/examples/下的 android 和 TensorFlow Mobile 的应用例子有很深的渊源，tensorflow/lite/java/下的 demo 一开始是作为 Android 的 Java 的应用例子出现的。两个文件夹里的活跃度都不低，现在还在保持更新。

另外，代码的相似度很高，使用的模型也很相似。现在还是不太清楚以后的这两个文件夹的发展计划。读者可以把两个文件夹都作为参考。

现在，我们来看一下 TensorFlow Lite 是怎样使用和集成模型文件的。tensorflow/java/demo/app/src/main/下的 build 文件定义了构建应用的方法，在 build 文件里，定义了我们要构建一个 tflite_demo 的应用，并且定义了在 assets 文件夹下面包含的文件。比如//tensorflow/examples/android/app/src/main/下的 assets:labels_mobilenet_quant_ v1_224.txt 是个文本文件，@tflite_mobilenet_quant//下的:mobilenet_v1_1.0_224_ quant.tflite 则是一个 TFLite 的模型文件。

```
    android_binary(
        name = "tflite_demo",
        srcs = glob([
            "app/src/main/java/**/*.java",
        ]),
        aapt_version = "aapt",
        assets = [
            "//tensorflow/lite/examples/android/app/src/main/assets:labels_
mobilenet_quant_v1_224.txt",
            "@tflite_mobilenet_quant//:mobilenet_v1_1.0_224_quant.tflite",
            "@tflite_conv_actions_frozen//:conv_actions_frozen.tflite",
            "//tensorflow/lite/examples/android/app/src/main/assets:conv_
actions_labels.txt",
            "@tflite_mobilenet_ssd//:mobilenet_ssd.tflite",
            "@tflite_mobilenet_ssd_quant//:detect.tflite",
            "//tensorflow/lite/examples/android/app/src/main/assets:box_
priors.txt",
            "//tensorflow/lite/examples/android/app/src/main/assets:coco_
labels_list.txt",
        ],
```

关于 mobilenet_v1_1.0_224_quant.tflite，在 tensorflow/workspace.bzl 里，可以看到这段代码：

```
    tf_http_archive(
        name = "tflite_mobilenet_quant",
        build_file = clean_dep("//third_party:tflite_mobilenet_quant.BUILD"),
        sha256  =  "d32432d28673a936b2d6281ab0600c71cf7226dfe4cdcef3012555f69
1744166",
        urls = [
            "http://download.tensorflow.org/models/mobilenet_v1_2018_08_
02/mobilenet_v1_1.0_224_quant.tgz",
            "http://download.tensorflow.org/models/mobilenet_v1_2018_08_
02/mobilenet_v1_1.0_224_quant.tgz",
        ],
    )
```

这段代码的意义是从 urls 指定的路径下载并解压缩文件，并用 tflite_mobilenet_quant.BUILD 去构建新的目标。下面是 tflite_mobilenet_quant.BUILD 的代码，它做的就是对于解压缩后的文件，忽略 build 文件，并把其他文件暴露出来，以便其他构建目标可以参照这些文件。这段代码里有两个相同的 URL，希望下个版本能够改正。

```
    exports_files(
        glob(
            ["**/*"],
            exclude = [
                "BUILD",
            ],
        ),
    )
```

如果下载 download.tensorflow.org/models/mobilenet_v1_2018_08_02/下的 mobilenet_v1_1.0_224_quant.tgz 并解压，我们可以得到

```
    -rw-r-----  1  17020468  Aug   2  18:38  mobilenet_v1_1.0_224_quant.ckpt.
data-00000-of-00001
    -rw-r----- 1    14644 Aug  2 18:38 mobilenet_v1_1.0_224_quant.ckpt.index
    -rw-r----- 1  5143394 Aug  2 18:38 mobilenet_v1_1.0_224_quant.ckpt.meta
    -rw-r----- 1   885850 Aug  2 18:38 mobilenet_v1_1.0_224_quant_eval.pbtxt
    -rw-r----- 1 17173742 Aug  2 18:38 mobilenet_v1_1.0_224_quant_frozen.pb
    -rw-r----- 1       89 Aug  2 18:38 mobilenet_v1_1.0_224_quant_info.txt
    -rw-r----- 1  4276352 Aug  2 18:39 mobilenet_v1_1.0_224_quant.tflite
    -rw-r----- 1 35069912 Aug  2 19:01 mobilenet_v1_1.0_224_quant.tgz
```

其中 mobilenet_v1_1.0_224_quant.tflite 就是应用中要使用的模型文件。

TensorFlow Lite Android 应用的代码和 TensorMobile 的代码在应用层面非常相似，比如：

- 读取摄像头的图像。
- 读取音频数据。
- 生成线程驱动模型。
- 存储模型。
- 读取模型。
- 用户界面。

两者的不同点主要是模型的不同和调用模型的方法不同，以及有些模型的输入和输出不同。我们在第 7 章讲解了应用层面的代码，以及调用模型的方法，这里就不重复了。下面，我们主要讲一下 TensorFlow Lite 的几个应用和它们的可视化模型视图。

先做些准备工作：

```
$ bazel build tensorflow/lite/tools:visualize
```

然后就可以使用构建后得到的 visualize 脚本工具，把 TensorFlow Lite 模型转换输出为 HTML 文件，文件包括很多信息，比如模型视图、Ops 等，我们下面会做详细说明。

7.3.1　声音识别

1. 模型

模型可以从 https://mirror.bazel.build/storage.googleapis.com/download.tensorflow.org/models/tflite/conv_actions_tflite.zip 下载，解压后可以看到如下文件：

```
-rw-rw-r-- 1 3771180 Mar  8  2018 conv_actions_frozen.tflite
-r--r--r-- 1      60 Mar  8  2018 conv_actions_labels.txt
-rw-r----- 1 3494186 Apr  2  2018 conv_actions_tflite.zip
```

使用上面提到的 visualize 脚本，运行下面的命令：

```
tensorflow/lite/tools/visualize conv_actions_frozen.tflite result.html
```

得到 result.html，然后在浏览器里加载这个文件，可以得到如图 7-5 所示的声音识别模型视图。

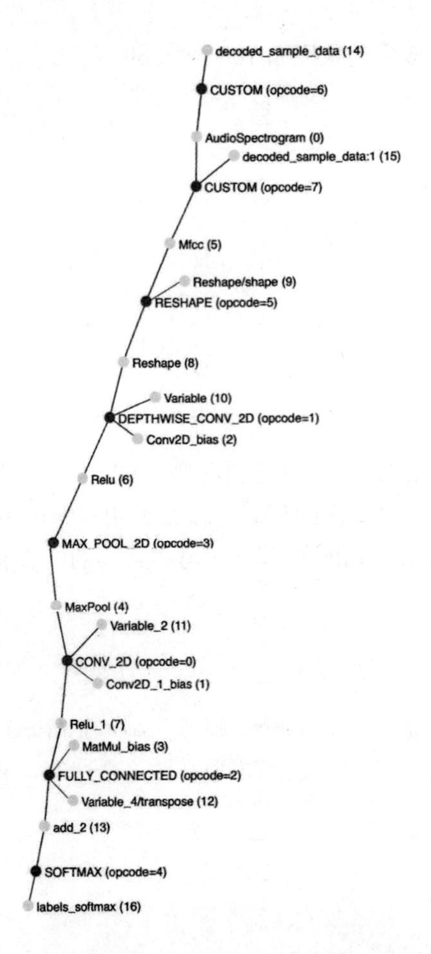

图 7-5　声音识别模型视图

输出的 HTML 文件里有很多信息，模型的概要信息（模型的名称、这个模型是如何得到的）如下：

```
filename conv_actions_frozen.tflite version 3 description TOCO Converted.
```

模型的输入输出的信息如下：

```
Subgraph 0
Inputs    Outputs
Inputs    outputs
[14, 15   [16]
```

从上面这些信息可以看到，输入是 14 和 15 张量，输出是 16 张量，这些数字对应的信息如表 7-1 所示。

表 7-1　Tensor

index	name	type	shape	buffer	quantization
0	AudioSpectrogram	FLOAT32	[]	12	{details_type: 0, quantized_dimension: 0}
1	Conv2D_1_bias	FLOAT32	[64]	2	{details_type: 0, quantized_dimension: 0}
2	Conv2D_bias	FLOAT32	[64]	8	{details_type: 0, quantized_dimension: 0}
3	MatMul_bias	FLOAT32	[12]	9	{details_type: 0, quantized_dimension: 0}
4	MaxPool	FLOAT32	[]	17	{details_type: 0, quantized_dimension: 0}
5	Mfcc	FLOAT32	[]	13	{details_type: 0, quantized_dimension: 0}
6	Relu	FLOAT32	[]	14	{details_type: 0, quantized_dimension: 0}
7	Relu_1	FLOAT32	[]	5	{details_type: 0, quantized_dimension: 0}
8	Reshape	FLOAT32	[]	15	{details_type: 0, quantized_dimension: 0}
9	Reshape/shape	INT32	[4]	16	{details_type: 0, quantized_dimension: 0}
10	Variable	FLOAT32	[1, 20, 8, 64]	7	{details_type: 0, quantized_dimension: 0}
11	Variable_2	FLOAT32	[64, 10, 4, 64]	6	{details_type: 0, quantized_dimension: 0}
12	Variable_4/transpose	FLOAT32	[12, 64000]	1	{details_type: 0, quantized_dimension: 0}
13	add_2	FLOAT32	[]	4	{details_type: 0, quantized_dimension: 0}
14	decoded_sample_data	FLOAT32	[16000, 1]	11	{details_type: 0, quantized_dimension: 0, min: [0.0], max: [255.0]}
15	decoded_sample_data:1	INT32	[1]	10	{details_type: 0, quantized_dimension: 0, min: [0.0], max: [255.0]}
16	labels_softmax	FLOAT32	[]	3	{details_type: 0, quantized_dimension: 0}

表 7-1 列出了所有的张量 Tensor 和对应的索引。如表 7-2 所示列出了所有的 Ops 和它们的输入输出的张量。

表 7-2　Ops

index	inputs	outputs	builtin_options	opcode_index
0	[14]	[0]	None	CUSTOM (opcode=6)
1	[0, 15]	[5]	None	CUSTOM (opcode=7)
2	[5, 9]	[8]	{new_shape: [-1, 99, 40, 1]}	RESHAPE (opcode=5)
3	[8, 10, 2]	[6]	{dilation_w_factor: 1, stride_h: 1, stride_w: 1, fused_activation_function: 'REL', depth_multiplier: 64, padding: 'SAME', dilation_h_factor: 1}	DEPTHWISE_CONV_2D (opcode=1)
4	[6]	[4]	{filter_height: 2, filter_width: 2, stride_h: 2, stride_w: 2, fused_activation_function: 'NONE', padding: 'SAME'}	MAX_POOL_2D (opcode=3)

（续表）

index	inputs	outputs	builtin_options	opcode_index
5	[4, 11, 1]	[7]	{dilation_w_factor: 1, stride_h: 1, stride_w: 1, fused_activation_function: 'REL', padding: 'SAME', dilation_h_factor: 1}	CONV_2D (opcode=0)
6	[7, 12, 3]	[13]	{weights_format: 'DEFAULT', fused_activation_function: 'NONE'}	FULLY_CONNECTED (opcode=2)
7	[13]	[16]	{beta: 1.0}	SOFTMAX (opcode=4)

如表 7-3 所示列出了每个张量所需存储的大小。

表 7-3　Buffer

index	data	index	data
0	--	9	48 bytes
1	3072000 bytes	10	--
2	256 bytes	11	--
3	--	12	--
4	--	13	--
5	--	14	--
6	655360 bytes	15	--
7	40960 bytes	16	16 bytes
8	256 bytes	17	--

演算子的信息如表 7-4 所示。请注意 6 和 7 是定制的演算子。

表 7-4　Operator Code

index	builtin_code	custom_code
0	CONV_2D	None
1	DEPTHWISE_CONV_2D	None
2	FULLY_CONNECTED	None
3	MAX_POOL_2D	None
4	SOFTMAX	None
5	RESHAPE	None
6	CUSTOM	AudioSpectrogram
7	CUSTOM	Mfcc

7.3.2　图像识别

前面的应用使用了 SSD Mobilenet 模型，该模型非常大，不能在书里全部展示，取出其中的一部分如图 7-6 所示，供读者参考。

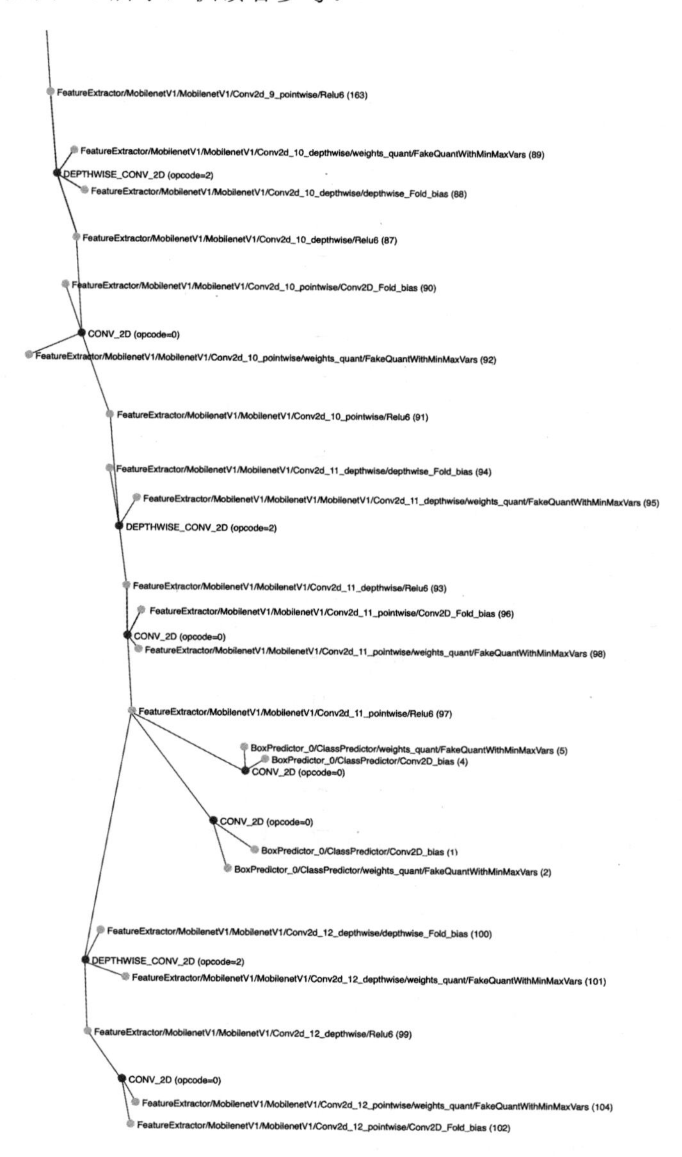

图 7-6　图像识别模型视图

由于模型数据非常大，详细内容就不写进本书了，有兴趣的读者可以尝试运行一下 Visualize 工具，自行查看视图。

7.4 TensorFlow Lite 使用 GPU

由于移动设备的处理能力有限，在移动设备上运行以计算密集型为主的机器学习模型的推理，对计算资源的要求很高。虽然转换为定点模型是加速的一种途径，但是运算效能还是不高。TensorFlow Lite 近期发布了对 GPU 的支持，使用 GPU 不会对程序产生额外的复杂性和潜在的量化精度损失。

TensorFlow Lite 对 GPU 的支持是通过 OpenGL ES 3.1 计算着色器（Compute Shaders）来实现的。读者可以下载并尝试使用 GPU 的预编译二进制预览版，TensorFlow Lite 官方许诺接下来会发布一个完整的开源版本。目前很多 TensorFlow Lite 的应用仍使用 CPU 进行运算，比如人脸轮廓检测，未来我们会利用新的 GPU 后端，预计可将 Pixel3 和三星 S9 的推理速度提升 4~6 倍。

7.4.1 GPU 与 CPU 性能比较

谷歌公布了一些他们的测试结果，这些结果是在谷歌的产品中进行几个月的测试后得到的。对于 Pixel3 的肖像模式，TensorFlow Lite GPU 让前景-背景分割模型的速度提高了 4 倍以上，新的深度预估模型的推理速度提高了 10 倍以上。在 YouTube 上的 YouTube Stories 和 Playground Stickers 中，实时视频分割模型在各种手机上的速度提高了 5~10 倍。对于各种各样的深度神经网络模型，新的 GPU 后端通常比浮点 CPU 实现速度快 2~7 倍。

7.4.2 开发 GPU 代理（Delegate）

GPU 旨在为大规模可并行化的计算工作负载提供高吞吐量。因此，GPU 非常适用于深度神经网络，深度神经网络由大量运算符组成，每个运算符处理一些输入张量，这些输入张量可以容易地分成较小的工作单元并且并行执行，这样做通常可以提高计算的延迟。在最佳情况下，GPU 上的推理运算现在足够快，可用于以前不可能实现的实时应用程序。

与 CPU 不同，GPU 使用 16 位或 32 位浮点数进行计算，并且不需要量化以获得最佳性能。GPU 推理的另一个好处是它的功效。GPU 以非常有效和优化的方式执行计算，因此与在 CPU 上运行相同任务相比，它们消耗更少的功率并产生更少的热量。

1. 演示应用程序

现在我们尝试在安卓上开发和集成支持 GPU 的 TensorFlow Lite 的预览版本。

这里我们使用 Gradle 来构建，有兴趣的读者可以尝试使用 Bazel 的 BUILD 来集成构建。具体实现代码如下：

```
dependencies {
    implementation 'org.tensorflow:tensorflow-lite:0.0.0-gpu-experimental'
}
```

接下来，在应用里添加 GPU 的支持代码，代码很简单，就是生成一个 GpuDelegate：

```
import org.tensorflow.lite.Interpreter;
import org.tensorflow.lite.experimental.GpuDelegate;

delegate = new GpuDelegate();
Interpreter.Options options = (new Interpreter.Options()).addDelegate
(delegate);
Interpreter interpreter = new Interpreter(model, options);

// 运行推理
while (true) {
  writeToInput(input);
  interpreter.run(input, output);
  readFromOutput(output);
}

// 清理空间
delegate.close();
```

2. GPU 支持的模型和算子

下面是 GPU 现在支持的模型的列表：

- MobileNet v1（224x224）图像分类器。它是为移动设备和嵌入式设备而设计的图像分类模型。

- DeepLab Segmentation（257x257）图像分割模型，将语义标签（例如狗、猫、汽车）分配给输入图像中的每个像素。
- MobileNet SSD Object Detection 图像分类模型，用于检测带有边界框的多个物体。
- PoseNet for Pose Estimation 视觉模型，用于估计图像或视频。

GPU 上的 TFLite 支持 16 位和 32 位浮点精度的如下操作：

- ADD
- AVERAGE_POOL_2D
- CONCATENATION
- CONV_2D
- DEPTHWISE_CONV_2D
- FULLY_CONNECTED
- LOGISTIC
- LSTM v2 (Basic LSTM only)
- MAX_POOL_2D
- MUL
- PAD
- PRELU
- RELU
- RELU6
- RESHAPE
- RESIZE_BILINEAR
- SOFTMAX
- STRIDED_SLICE
- SUB
- TRANSPOSE_CONV

3. 不支持的模型和操作

如果 GPU 的代理不支持某些操作，则 TFLite 将仅在 GPU 上运行图形中可支持的 Ops，而在 CPU 上运行其余部分。由于 CPU/GPU 同步的成本高，像这样的拆分执行模式通常会导致性能比单独在 CPU 上运行整个网络时慢。在这种情况下，程序会产生如下警告：

```
WARNING: op code #42 cannot be handled by this delegate.
```

程序没有为此失败提供回调函数，因为这不是真正的运行时错误，而是开发人员在尝试让网络在 GPU 代理上运行时可以观察到的内容。

4. 实现原理

深度神经网络会按顺序运行数百个操作，使它们非常适合 GPU，GPU 的设计考虑了面向高吞吐量的并行工作负载。

Interpreter::ModifyGraphWithDelegate() 在 C++ 中调用时初始化 GPU，或者通过 Interpreter.Options 间接调用 Interpreter 的构造函数来初始化 GPU。下面是用 C++实现的参考代码：

```
auto model = FlatBufferModel::BuildFromFile(model_path);
tflite::ops::builtin::BuiltinOpResolver op_resolver;
std::unique_ptr<Interpreter> interpreter;
InterpreterBuilder(*model, op_resolver)(&interpreter);

auto* delegate = NewGpuDelegate(nullptr);  // default config
QCHECK_EQ(interpreter->ModifyGraphWithDelegate(delegate), kTfLiteOk);

WriteToInputTensor(interpreter->typed_input_tensor<float>(0));
QCHECK_EQ(interpreter->Invoke(), kTfLiteOk);
ReadFromOutputTensor(interpreter->typed_output_tensor<float>(0));

DeleteGpuDelegate(delegate);
```

在初始化阶段，基于框架接收的执行计划（Execution Plan）构建输入神经网络的表示。使用这种新的表示，可以实现一些有用的转换，这些转换包括如下内容：

- 剔除不需要的 Ops。
- 将 Ops 替换为性能更好的等效 Ops。

- 合并 Ops，以减少最终生成的着色程序的数量。

然后基于此优化图（Optimized Graph），生成并编译计算着色器。在 Android 上现在使用 OpenGL ES 3.1 计算着色器，在创建这些计算着色器时，程序还采用了各种结构的优化，例如：

- 应用某些操作的特化而不是它们的通用实现。
- 释放寄存器压力。
- 选择最佳工作组大小。
- 安全地调整精度。
- 重新排序显式数学操作。

在这些优化结束时，着色器程序被编译，可能需要几毫秒到半秒，就像手机游戏一样。一旦着色程序编译完成，新的 GPU 推理引擎就可以开始工作了。

在对每个输入进行推理运算时：

- 如有必要，输入将移至 GPU。输入张量（如果尚未存储为 GPU 内存）可由框架通过创建 GL 缓冲区或 MTLBuffers 进行 GPU 访问，同时还可能复制数据。由于 GPU 在 4 通道数据结构中效率最高，因此通道大小不等于 4 的张量将重新调整为更加适合 GPU 的架构。
- 执行着色器程序。将上述着色器程序插入命令缓冲区队列中，GPU 将这些程序输出。在此步骤中，我们还为中间张量管理 GPU 内存，以尽可能减少后端的内存占用。
- 必要时将输出移动到 CPU。一旦深度神经网络处理完成，框架将结果从 GPU 内存复制到 CPU 内存，除非网络的输出可以直接在屏幕上呈现，不需要这样的传输。

7.5 训练模型

TensorFlow Lite 主要是面向推理（Inference）的，但是，随着硬件设备的发展和技术的进步，未来在移动设备上进行模型训练并商业化，应该也不是不可能的事。我们在这里介绍一个简单实现的代码。

7.5.1　仿真器

首先，准备仿真器（Emulator）。在安装 Android SDK 时，请同时安装 Intel x86 Atom System 映像。

然后，让我们列出可用的 Android 模拟器，AVD 是 Android Virtual Devices 的简称。

```
$ /opt/Android/sdk/tools/android list avd
*****************************************************************************
The "android" command is deprecated.
For manual SDK, AVD, and project management, please use Android Studio.
For command-line tools, use tools/bin/sdkmanager and tools/bin/avdmanager
*****************************************************************************
Running /opt/Android/sdk/tools/bin/avdmanager list avd
```

或，执行下面的命令：

```
$ /opt/Android/sdk/tools/bin/avdmanager list avd

Available Android Virtual Devices:
    Name: Pixel_API_26
  Device: pixel (Google)
    Path: .android/avd/Pixel_API_26.avd
  Target: Google APIs (Google Inc.)
          Based on: Android 8.0 (Oreo) Tag/ABI: google_apis/x86
    Skin: pixel
  Sdcard: 100M
```

现在，我们可以选择一个模拟器并运行它，比如执行下面的命令：

```
/opt/Android/sdk/tools/emulator -avd Pixel_API_26
```

7.5.2　构建执行文件

在 tensorflow/cc/build 中，添加以下目标：

```
cc_binary(
    name = "android_tutorials_example_trainer",
    srcs = ["tutorials/example_trainer.cc"],
    copts = tf_copts(),
    linkopts = select({
```

```
            "//tensorflow:android": [
                "-lm",
                "-fPIE",
                "-pie",
                "-llog",
                "-latomic", # x86 requires
                "-landroid",
                "-Wl,-z,defs",
                "-Wl,--no-undefined",
                "-s",
            ],
            "//conditions:default": [
                "-lm",
                "-lpthread",
                "-lrt",
            ],
        }),
        deps = [
            "while_loop",
            "//tensorflow/core:android_tensorflow_lib",
        ],
    )
```

现在，我们可以通过下面的命令来构建执行文件：

```
    $ bazel build -c opt --cxxopt='--std=c++11' --crosstool_top=//external:
android/crosstool      --host_crosstool_top=@bazel_tools//tools/cpp:toolchain
--config=android --cpu=x86 //tensorflow/cc:android_tutorials_example_trainer
```

得到的执行文件如下：

```
    bazel-bin/tensorflow/cc/android_tutorials_example_trainer
```

为使我们有权将可执行文件复制到移动设备，执行如下命令：

```
    adb root
```

然后，复制文件并运行：

```
    $   adb   push   bazel-bin/tensorflow/cc/android_tutorials_example_trainer
/data/local/tmp/
```

最后，执行训练，执行命令及结果如下：

```
    $ adb shell /data/local/tmp/android_tutorials_example_trainer
    native : cpu_feature_guard.cc:35 The TensorFlow library was compiled to use
```

```
SSE instructions, but these aren't available on your machine.
    native : cpu_feature_guard.cc:35 The TensorFlow library was compiled to use
SSE2 instructions, but these aren't available on your machine.
    native : cpu_feature_guard.cc:35 The TensorFlow library was compiled to use
SSE3 instructions, but these aren't available on your machine.
    can't determine number of CPU cores: assuming 4
    can't determine number of CPU cores: assuming 4
    000000/000001 lambda = 3.056837 x = [0.969245 0.246098] y = [3.399931
-0.969245]
    000000/000003 lambda = 1.716893 x = [0.639100 0.769124] y = [3.455547
-0.639100]
    000000/000003 lambda = 2.721944 x = [0.983324 -0.181865] y = [2.586241
-0.983324]
    000000/000003 lambda = 2.288898 x = [0.934717 -0.355392] y = [2.093368
-0.934717]
    000000/000003 lambda = 2.129015 x = [0.913109 -0.407716] y = [1.923896
-0.913109]
    000000/000003 lambda = 2.061103 x = [0.903412 -0.428773] y = [1.852692
-0.903412]
    000000/000003 lambda = 2.029754 x = [0.898833 -0.438291] y = [1.819919
-0.898833]
    000000/000003 lambda = 2.014684 x = [0.896609 -0.442823] y = [1.804181
-0.896609]
    000000/000003 lambda = 2.007294 x = [0.895513 -0.445036] y = [1.796467
-0.895513]
    000000/000003 lambda = 2.003635 x = [0.894969 -0.446129] y = [1.792648
-0.894969]
    000000/000003 lambda = 2.001815 x = [0.894698 -0.446672] y = [1.790748
-0.894698]
    000000/000003 lambda = 2.000907 x = [0.894562 -0.446943] y = [1.789801
-0.894562]
    000000/000003 lambda = 2.000453 x = [0.894495 -0.447078] y = [1.789327
-0.894495]
    000000/000003 lambda = 2.000227 x = [0.894461 -0.447146] y = [1.789091
-0.894461]
    ... ... ...
    000000/000005 lambda = 2.000000 x = [0.894427 -0.447214] y = [1.788854
-0.894427]
    000000/000005 lambda = 2.000000 x = [0.894427 -0.447214] y = [1.788854
-0.894427]
    000000/000005 lambda = 2.000000 x = [0.894427 -0.447214] y = [1.788854
-0.894427]
```

8

第 8 章 移动端的机器学习开发

本章前半部分简要介绍 TensorFlow 对其他移动平台和嵌入式平台的支持，后半部分介绍移动端 TensorFlow 开发的实战经验。

8.1 其他设备的支持

下面介绍 TensorFlow 对 iOS 和树莓派的支持。iOS 在移动端占了很大份额，是每个开发人员都会关注的平台。树莓派作为 IoT 和小型设备的代表，在其上面进行移动开发有很重要的示范意义。

8.1.1 在 iOS 上运行 TensorFlow 的应用

我们已经在 iOS 上看到了很多使用 TensorFlow 的优秀的应用程序，iOS 大概占有 20%

的市场份额，所以支持这个平台对 TensorFlow Lite 很重要。如果读者需要进行开发，TensorFlow 提供了完整的从源代码构建到开发应用的方法。

1. 使用 CocoaPods

在 iOS 上使用 TensorFlow 最简单的方法是使用 CocoaPods 包管理系统。读者可以从 cocoapods.org 上下载 TensorFlow 框架包，然后只需运行“TensorFlow-experimental”，将其作为依赖添加到你的应用程序的 Xcode 项目中。这样就安装了一个通用的二进制框架，这使得开发者很容易入门，但缺点是开发者难以定制软件包，这对于缩小应用的二进制大小非常重要。

开发者可以执行下面的脚本，完成安装。

```
tensorflow/contrib/makefile/build_all_ios.sh
```

这个过程大约需要 20 分钟。

2. 使用 Makefile Unix

make 程序可能是最古老的实用构建工具之一，用 make 直接构建和操作文件，这种较底层的操作方式为一些棘手的开发情况提供了很多的灵活性，比如在交叉编译或在旧的及有限的资源系统上构建时，TensorFlow 在 tensorflow/contrib/makefile 中提供了一个针对移动和嵌入式平台的 Makefile。这是构建 iOS 的主要方式，针对 Linux、Android 和树莓派也是有用的。下面介绍一下使用这种方法进行手动构建的过程。

首先，执行下面的脚本，下载依赖的软件包：

```
tensorflow/contrib/makefile/download_dependencies.sh
```

然后，执行下面的脚本为 iOS 编译 ProtoBuf：

```
tensorflow/contrib/makefile/compile_ios_protobuf.sh
```

接着，运行指定 iOS 作为目标的 Makefile，以及 iOS 相对应的芯片架构：

```
make -f tensorflow/contrib/makefile/Makefile \
    TARGET=IOS \
    IOS_ARCH=ARM64
```

执行结束后，会生成 tensorflow/contrib/makefile/gen/lib/libtensorflow-core.a 的一个通用软件库，开发人员可以使用这个库链接任何 Xcode 项目。

3. 编译优化

compile_ios_tensorflow.sh 脚本可以接受可选的命令行参数。第一个参数是 C++优化参数，默认为调试模式。开发者可以配置更高的优化参数，如下所示：

```
compile_ios_tensorflow.sh "-Os"
```

有关优化参数的其他选项，请参见优化级别的文档。

iOS 示例有三个演示应用程序，全部定义在 tensorflow/contrib/ios_examples 中的 Xcode 项目中。

4. 简单演示

演示如何用最少的代码加载和运行 TensorFlow 模型的最简单示例。它由一个单一的视图和一个按钮组成，当按下按钮时，会执行模型加载和推理。

5. 相机演示

类似于 Android 的 TensorFlow 图像分类演示。它加载了 Inception v3，并输出其最佳图像标签，也可以实时查看相机视图中的内容。与 Android 版本一样，开发者可以使用 TensorFlow for Poets（https://codelabs.developers.google.com/codelabs/tensorflow-for-poets）来训练开发者自定义的模型，并将其放入此示例中，只需更改少量代码即可。

6. 性能测试演示

与简单演示很接近，但它重复运行模型，并将统计数据输出给基准测试（Benchmark）工具。要构建这些演示，首先要确保你已经能够成功地为 iOS 编译主 TensorFlow 库。开发者还需要下载需要的模型文件，请运行下面的代码：

```
mkdir -p ~/graphs
curl -o ~/graphs/inception5h.zip \
    https://storage.googleapis.com/download.tensorflow.org/models/
inception5h.zip
    unzip ~/graphs/inception5h.zip -d ~/graphs/inception5h
cp ~/graphs/inception5h/* \
    tensorflow/examples/ios/benchmark/data/
cp ~/graphs/inception5h/* \
    tensorflow/examples/ios/camera/data/
```

```
cp ~/graphs/inception5h/* \
    tensorflow/examples/ios/simple/data
```

开发者应该能够为每个单独的演示加载 Xcode 项目，构建它并在设备上运行。相机演示需要一个相机，所以它不会在模拟器上运行，但图像渲染和声音识别应用应该可以在模拟器上运行。开发者也可以直接从 iOS 应用程序直接使用 C++，代码可以直接调用 TensorFlow 框架的 API。

8.1.2　在树莓派上运行 TensorFlow

TensorFlow 团队正在努力为树莓派提供一个官方的集成包的安装路径，以便使用预先构建的二进制文件在树莓派上运行该框架。在写这篇文章的时候它还不能用（查看 https://www.tensorflow.org/install 了解最新的细节），在这里笔者将介绍如何从源代码构建它。

基于树莓派的构建与普通的 Linux 系统类似。首先，下载并安装依赖的软件包，构建 ProtoBuf：

```
tensorflow/contrib/makefile/download_dependencies.sh
sudo  apt-get install -y \
autoconf automake libtool gcc-4.8 g++-4.8
cd tensorflow/contrib/makefile/downloads/protobuf/
./autogen.sh
./configure
make
sudo make install
sudo ldconfig # refresh shared library cache
cd ../../../../..
```

然后，使用 make 命令来建立库和应用例子：

```
make -f tensorflow/contrib/makefile/Makefile HOST_OS=PI \
     TARGET=PI OPTFLAGS="-Os -mfpu=neon-vfpv4 \
      -funsafe-math-optimizations -ftree-vectorize" CXX=g++-4.8
```

下面介绍树莓派的应用例子。

树莓派是各种嵌入式应用原型的绝佳平台。在 tensorflow/contrib/pi_examples 中有两个不同的例子：

1. 图像标签

图像标签是标准的 tensorflow/examples/label_image 演示的一个移植演示，它尝试利用 Inception v3 Imagenet 的模型来标记图像。与其他平台一样，开发者可以轻松地使用从 TensorFlow for Poets 派生的定制训练版本来替换此模型。

2. 相机

这个例子使用树莓派的摄像头 API 来读取一个实时视频，在其上运行图像标签，并将标签输出到控制台。为了让演示更有趣，此应用可以让开发者将结果提供给 tflite 文本语音工具，以便让树莓派能说出所看到的内容。要构建这些示例，请确保你已经运行了前面所示的树莓派构建过程，然后执行如下命令：

```
$ makefile -f tensorflow/contrib/pi_examples/camera
```

或执行如下命令：

```
$ makefile -f tensorflow/contrib/pi_examples/simple
```

执行完命令将生成一个根文件夹 gen/bin，运行其中的可执行文件即可获取模型文件，示例代码如下：

```
curl https://storage.googleapis.com/download.tensorflow.org/ \
      models/inception_dec_2015_stripped.zip \
      -o /tmp/inception_dec_2015_stripped.zip
   unzip /tmp/inception_dec_2015_stripped.zip \
      -d tensorflow/contrib/pi_examples/label_image/data/
```

8.2 设计和优化模型

以上我们学习和研究了 TensorFlow Mobile 和 TensorFlow Lite，并大致了解了 iOS 和树莓派的开发，下面我们总结一下如何在移动端和嵌入式设备上进行机器学习的开发，以及开发中需要注意的地方。当开发者尝试在移动设备或嵌入式设备上发布应用时，有一些特殊问题需要处理，比如怎样优化延迟，优化内存（RAM）的使用，优化模型文件大小和执行文件二进制大小，开发者在开发模型时也需要考虑这些问题。

8.2.1　模型大小

模型需要存储在设备上的某个地方，大型的神经网络可能需要数百兆字节的存储空间。即使有足够的本地存储空间，当用户需要从应用商店下载非常大的应用程序包时，就会占据很多存储空间，因此开发者需要规划模型的规模。

在使用 freeze_graph 和 strip_unused_nodes 之后，开发者可以查看 GraphDef 文件在磁盘上的大小，因为它应该只包含推理相关的节点。要仔细检查模型是否符合预期，开发者可以通过 summarize_graph 脚本文件来查看常量中有多少个参数：

```
bazel build tensorflow/tools/graph_transforms:summarize_graph \
 && bazel-bin/tensorflow/tools/graph_transforms/summarize_graph \
 --in_graph=/tmp/tensorflow_inception_graph.pb
```

代码执行结果如下：

```
No inputs spotted.
Found 1 possible outputs: (name=softmax, op=Softmax)
Found 23885411 (23.89M) const parameters, 0 (0) variable
 parameters, and 99 control_edges
Op types used: 489 Const, 99 CheckNumerics, 99 Identity,
 94 BatchNormWithGlobalNormalization, 94 Conv2D, 94 Relu,
 11 Concat, 9 AvgPool, 5 MaxPool, 1 Sub, 1 Softmax,
 1 ResizeBilinear, 1 Reshape, 1 Mul, 1 MatMul, 1 ExpandDims,
 1 DecodeJpeg, 1 Cast, 1 BiasAdd
```

注意，结果中常量参数的个数是 23 885 411，一般每个常量参数是以 32 位浮点数存储的，我们用参赛个数乘以 4，得到的数值应该和模型文件的大小比较接近。如果我们把 32 位数值改成 8 位数值，那么模型的损失会非常小，可以省下很多的存储空间。如果模型过大，开发者可以采用这种方法减少模型的大小。具体实现代码如下：

```
bazel build tensorflow/tools/graph_transforms:transform_graph && \
bazel-bin/tensorflow/tools/graph_transforms/transform_graph \
--in_graph=/tmp/tensorflow_inception_optimized.pb \
--out_graph=/tmp/tensorflow_inception_quantized.pb \
--inputs='Mul:0' --outputs='softmax:0' --transforms='quantize_weights'
```

注： 参数中的“--transforms='quantize_weights'”表示使用定点化的方法。

另外一种模型压缩方法是 round_weights。这种方法不能压缩模型本身的大小，但是可以使模型在压缩后的文件更小。这种方法将权重参数保存为浮点数，但将其舍入为设定的步数值。这意味着存储模型中有更多的重复字节模式，所以压缩常常会大大降低文件的大小。

在很多情况下，最终的文件大小可能会非常接近 8 位模型。这样做的优点是，框架不必分配一个临时缓冲区来解压缩参数，就像我们在使用 quantize_weights 时一样。这样可以适当节省运行的延迟。

我们还可以使用内存映射，以此减少加载模型的时间，这在本书的第 7 章中讲过。

8.2.2 运行速度

大多数模型开发和部署的最高优先级之一就是，如何快速运行推理以提供良好的用户体验。这个过程的第一步是，查看执行图所需的浮点运算的总数。开发者可以通过使用 benchmark_model 脚本工具来得到一个非常粗略的估算，代码如下：

```
bazel build -c opt tensorflow/tools/benchmark:benchmark_model \
 && bazel-bin/tensorflow/tools/benchmark/benchmark_model \
 --graph=/tmp/inception_graph.pb --input_layer="Mul:0" \
 --input_layer_shape="1,299,299,3" --input_layer_type="float" \
 --output_layer="softmax:0" \
 --show_run_order=false --show_time=false \
 --show_memory=false --show_summary=true --show_flops=true \
 --logtostderr
```

运行这个脚本之后，就会显示需要多少操作来运行 TensorFlow 模型图。然后开发者就可以通过这些信息来确定模型在目标设备上运行的可行性。举个例子，2016 年的高端手机每秒钟可以处理 200 亿个 FLOP，所以运行一个需要 100 亿个 FLOP 的模型的最佳速度大约是 500 毫秒。在像树莓派 3 那样可以做大约 50 亿个 FLOP 的设备上，大概每两秒钟只能得到一个推理结果。

有了这个估计之后就可以规划在移动设备上可以实现的功能，如果模型使用的操作太多，那么我们就要尽量优化架构以减少这个运算的数量。另外，也可以选择比较先进的模型，比如 SqueezeNet 和 MobileNet。也可以寻找替代模型，甚至可能更小的旧模型。例如，Inception v1 只有约 700 万个参数，运行需要 90 亿个 FLOP。而 Inception v3 的 2400 万个参数运行需要 30 亿个 FLOP。

如何检验你的模型？

一旦开发者了解了移动设备可能达到的最佳性能，我们就要看看应用能获得的实际性能。我们推荐在比较独立和隔绝的空间里运行应用，不推荐和其他应用程序混合运行，因为这有助于隔离 TensorFlow 对延迟的影响。TensorFlow 的基准工具（Benchmark）可以帮助开发者做到这一点。如果要检验 Inception v3 的性能，请运行如下代码：

```
bazel build -c opt tensorflow/tools/benchmark:benchmark_model \
 && bazel-bin/tensorflow/tools/benchmark/benchmark_model \
 --graph=/tmp/tensorflow_inception_graph.pb \
 --input_layer="Mul" --input_layer_shape="1,299,299,3" \
 --input_layer_type="float" --output_layer="softmax:0" \
 --show_run_order=false --show_time=false \
 --show_memory=false --show_summary=true \
 --show_flops=true --logtostderr
```

输出结果如下：

```
    ======================== Top by Computation Time =========================
[node type]    [start]    [first]    [avg ms]    [ % ]    [cdf%]    [mem KB]    [Name]
Conv2D         22.859     14.212     13.700      4.972%   4.972%    3871.488    conv_4/
                                                                                Conv2D
Conv2D         8.116      8.964      11.315      4.106    9.078%    5531.904    conv_2/
                                                                                Conv2D
Conv2D         62.066     16.504     7.274       2.640%   11.717%   443.904     mixed_3/
                                                                                conv/
                                                                                Conv2D
Conv2D         2.530      6.226      4.939       1.792%   13.510%   2765.952    conv_1/
                                                                                Conv2D
Conv2D         55.585     4.605      4.665       1.693%   15.203%   313.600     mixed_2/
                                                                                tower/
                                                                                conv_1/
                                                                                Conv2D
Conv2D         127.114    5.469      4.630       1.680%   16.883%   81.920      mixed_10/
                                                                                conv/
                                                                                Conv2D
Conv2D         47.391     6.994      4.588       1.665%   18.548%   313.600     mixed_1/
                                                                                tower/
                                                                                conv_1/
                                                                                Conv2D
Conv2D         39.463     7.878      4.336       1.574%   20.122%   313.600     mixed/
                                                                                tower/
                                                                                conv_1/
                                                                                Conv2D
Conv2D         127.113    4.192      3.894       1.413%   21.535%   114.688     mixed_10/
                                                                                tower_1/
                                                                                conv/
```

```
                                                                            Conv2D
Conv2D         70.188     5.205      3.626       1.316%  22.850%  221.952  mixed_4/
                                                                              conv/
                                                                            Conv2D
    ======================== Summary by node type ==========================
    [Node type]      [count]      [avg ms]       [avg %]       [cdf%]        [mem KB]
    Conv2D              94        244.899        88.952%       88.952%       35869.953
    BiasAdd             95        9.664          3.510%        92.462%       35873.984
    AvgPool              9        7.990          2.902%        95.364%       7493.504
    Relu                94        5.727          2.080%        97.444%       35869.953
    MaxPool              5        3.485          1.266%        98.710%       3358.848
    Const              192        1.727          0.627%        99.337%       0.000
    Concat              11        1.081          0.393%        99.730%       9892.096
    MatMul               1        0.665          0.242%        99.971%       4.032
    Softmax              1        0.040          0.015%        99.986%       4.032
    <>                   1        0.032          0.012%        99.997%       0.000
    Reshape              1        0.007          0.003%        100.000%      0.000

    Timings (microseconds): count=50 first=330849 curr=274803 min=232354
    max=415352 avg=275563 std=44193
    Memory (bytes): count=50 curr=128366400(all same)
    514 nodes defined 504 nodes observed
```

上面是输出的摘要，注意脚本中 show_summary 的参数应设为真。输出中第一个表格是按时间顺序花费最多时间的节点列表。从左到右依次是：

- 节点类型，即什么样的操作。
- 操作的开始时间。
- 第一次运算的时间，以毫秒为单位。这是基准测试第一次运行需要多长时间，默认情况下会执行 20 次运行以获得更可靠的统计数据。可以发现在第一次运行中进行较长的计算并且缓存结果的操作是有用的。
- 所有运行的平均运行时间，以毫秒为单位。
- 运行一次所花时间的百分比。这对了解主要计算的发生位置非常有用。
- 表中这个操作和以前操作的累计总时间。这对于理解神经网络层间的工作分布是很方便的，看看是否有少数节点在大部分时间都在被使用。
- 节点的名称。

第二个表与第一个表类似，但不是按特定的运算节点来划分时间，而是按照操作类型进行分组。这对想要了解从图形中优化或消除哪些操作非常有用。该表只显示前 10 个类型分类，从左到右依次是：

- 正在分析的节点的类型。
- 此类型的所有节点累积的平均时间，以毫秒为单位。
- 这种类型的操作占总时间的百分比。
- 表中这个节点类型的累计时间较长，因此开发者可以了解工作负载的分布情况。
- 此节点类型的输出占用了多少内存。

两个表格都有作为列之间分隔符的选项符号，因此可以轻松地将结果复制并粘贴到电子表格中。在查找可以优化的点时，总结节点类型可能是最有用的，因为它会告诉开发者哪些操作花费时间最多。在这种情况下，我们一般可以看到 Conv2D 的运行时间几乎是执行时间的 90%。该迹象表明该图是经过良好优化的，因为卷积和矩阵乘法计算预计占了神经网络的计算工作量的大部分。

作为一个经验法则，如果你看到其他操作占用了很多时间，那是不太好的迹象。对于神经网络，不涉及大矩阵乘法的操作通常应该作为等级低的操作被处理。因此，如果开发者看到很多时间花费在这些操作上，则表明这个神经网络不是最佳结构，或者实现这个神经网络的操作代码没有被优化。遇到这种情况，开发者要尽量做性能缺陷修补程序。特别是，当你使用基准测试（Benchmark）工具看到了相似的问题时，更要注意。

上面的工具是运行在开发者的桌面电脑上的，但这些工具也可以运行在 Android 上，其实这是最适合移动开发的环境。以下是在 64 位 ARM 设备上运行它的示例代码：

```
    bazel build -c opt --config=android_arm64 \
         tensorflow/tools/benchmark:benchmark_model
    adb push bazel-bin/tensorflow/tools/benchmark/benchmark_model/data/local/
tmp
    adb push /tmp/tensorflow_inception_graph.pb /data/local/tmp/
    adb shell '/data/local/tmp/benchmark_model \
       --graph=/data/local/tmp/tensorflow_inception_graph.pb \
       --input_layer="Mul" --input_layer_shape="1,299,299,3" \
       --input_layer_type="float" --output_layer="softmax:0" \
       --show_run_order=false --show_time=false
       --show_memory=false --show_summary=true'
```

对于 iOS 上可以执行的命令行工具，现在还没有很好的支持，所以在 tensorflow/contrib/ios_examples/benchmark 中有一个单独的例子，它将应用程序中的相同功能打包。这会将统计信息输出到设备屏幕并进行日志调试。

8.2.3 可视化模型

加速开发者的代码最有效的方法是改变你的模型，它可以减少模型的计算量。要做到这一点，开发者需要了解模型正在做什么，并将其作为可视化的第一步。

为了高度概括模型的运行图，建议开发者使用 Tensor Board 应用程序。它应该能够加载开发者的 GraphDef 文件，虽然有时可能会遇到较大的运行图。为了获取更细粒度的视图，我们可以将图形转换为 Graphviz 的 DOT 文件格式。下面是相关的脚本代码：

```
bazel build tensorflow/tools/quantization:graph_to_dot
bazel-bin/tensorflow/tools/quantization/graph_to_dot \
    --graph=/tmp/tensorflow_inception_graph.pb \
    --dot_output=/tmp/tensorflow_inception_graph.dot
```

如果读者使用的是类 Unix 环境，可以安装 xdot 程序。运行 xdot 可能需要花费几秒钟或更长的时间来加载非常大的运行图。在默认情况下，脚本只显示操作类型，因为名称可能很长，屏幕输出很难看。当然，如果需要，也可以修改 Python 脚本以显示更多信息。

8.2.4 线程

TensorFlow 的桌面版本具有复杂的线程模型，如果可以，将尝试并行运行多个操作，术语叫作“内部操作并行”。可以通过在会话（Session）选项中指定“interop_threads”来设置操作间的并行性。

在移动设备上，系统内部操作并行处理数（一次运行多少个操作数）默认设置为 1，以便操作始终按顺序连续运行。这种设置是合理的，因为移动处理器通常具有很少的内核数和一个小的缓存，所以运行多个访问不相交部分内存的操作通常不会提高性能。操作内并行是非常有用的，比如对于卷积运算，在多线程运行的情况下，可以为每个线程分配一块小的内存。

在移动设备上，默认情况下，一个操作系统使用的线程数将被设置为核心数量，或者在核心数量无法确定的情况下，该数量将只有 4 个。读者可以通过使用会话选项显式地设置它来覆盖默认的线程数。如果你的应用程序有自己的线程处理池，那么它们就不会相互干扰，所以减少线程默认值是个好主意。

8.2.5　二进制文件大小

移动设备和桌面服务器开发之间最大的区别是二进制文件大小。在桌面计算机上，数百兆字节的大型可执行文件并不罕见，但对于移动和嵌入式应用程序来说，保持二进制文件尽可能小是至关重要的，这样用户的下载将变得非常简单和快捷。

TensorFlow 的应用默认只包含 TensorFlowOps 的一个子集，但是这样仍然导致最终可执行文件非常大。为了减少文件的大小，开发者可以设置自动分析模型来设置库，只包含应用实际需要的 Ops 的实现。要使用这种方法，请按照下列步骤进行操作：

（1）在你的模型上运行

tools/print_required_ops/目录下的 print_selective_registration_header.py 脚本，产生一个只启用它所使用的操作的头文件。

（2）将 ops_to_register.h 文件放在编译器可以找到的地方，比如放在 TensorFlow 源文件夹的根目录下。

（3）使用 SELECTIVE_REGISTRATION 构建 TensorFlow，例如将 --copts ="-DSELECTIVE_REGISTRATION"传递给 Bazel 构建命令。

此过程会重新编译库，以便只包含所需操作和类型，这样可以显著减少可执行文件的大小，我们在开发中一定要有这个步骤。例如，在 Inception v3 中，新的模型文件大小只有 1.5MB。

8.2.6　重新训练移动数据

在移动应用上运行模型时，产生精度问题的最大原因是，使用了没有代表性的训练数据。例如，大多数的 Imagenet 照片都是精心设计的，所以物体位于照片的中心，光线充足，镜头正常。

移动设备拍摄的照片尤其自拍照，通常照相环境很差，照明不足，可能会造成鱼眼失真。解决方案是，扩展开发者应用程序中实际收集的数据。这一步可能需要额外的工作，因为开发者必须要自己进行标注。

虽然我们只是扩展了原始的训练数据，但是极大地提高了准确性。这比改变模型架构或使用不同的技术要有效得多。

8.2.7 优化模型加载

大多数操作系统允许开发者使用内存映射加载文件，而不是通过通常的输入输出API。我们不希望在内存堆上分配一个内存区域，将模型字节从磁盘复制到内存中，我们希望仅仅告诉操作系统文件的内容，让操作系统把模型映射到内存中。这种方法的优点是，操作系统知道整个文件将立即被读取，并且可以有效地规划加载过程，因此可以非常快。实际的加载也可以暂停，直到内存被第一次访问，所以它与代码的初始化是异步的。

开发者也可以告诉操作系统，你只能从内存区域读取数据，而不是写入数据。这样做的好处是，当 RAM 受到存储压力时，系统不会把数据从虚拟内存里保存到磁盘，而是可以放弃，因为磁盘上已经有了模型，我们可以再次加载，节省了大量的磁盘写入时间。

由于 TensorFlow 模型通常可能有几兆字节或更大，因此加快加载过程对于移动和嵌入式应用程序来说是一个很大的帮助，减少写入负载也可以对系统响应性有很大的帮助。

减少内存使用量也是非常有用的。例如，在 iOS 上，系统可以终止使用超过 100MB RAM 的应用程序，特别是在较旧的设备上。内存映射文件使用的内存不会计入这个限制，所以它通常是这些设备上模型的最佳选择。

构成 TensorFlow 模型的大部分是模型的权重。由于 Protobuf 序列化格式的限制，我们必须对模型加载和处理代码进行一些更改。内存映射的工作方式是，假如我们有一个单独的文件，其中第一部分是一个正常的 GraphDef，它串行化成协议缓冲区格式，权重可以通过直接映射的形式加载。要创建此文件，读者需要运行 tensorflow/contrib/目录下的 util:convert_graphdef_memmapped_format 工具。该工具接受一个已经通过 freeze_graph 运行的 GraphDef 文件，并将其转换为最后附加权重的格式。由于该文件不再是标准的 GraphDef Protobuf，因此需要对加载代码进行一些更改。读者可以在 iOS 相机演示应用程序 LoadMemoryMappedModel()函数中看到这个示例。

8.2.8 保护模型文件

在默认情况下，开发者的模型将以磁盘上的标准序列化 Protobuf 格式存储。理论上这意味着任何人都可以复制你的模型，所以我经常被问及如何防止这种情况。在实践中，大多数模型都是特定于应用程序的，并且通过优化来混淆和隐藏模型，可以避免竞争对手采用类似于拆解和反向工程的风险。如果你想让临时用户更难访问你的文件，可以使用下面的方法进行加密和解密。

大多数示例使用 ReadBinaryProto 简单地从磁盘加载 GraphDef。这一步需要读磁盘上的未加密的 Protobuf。幸运的是，调用的实现过程非常简单，编写一个可以在内存中解密的等价函数是很容易的。以下是一些代码，展示了如何使用自己的解密过程来读取和解密 Protobuf：

```
Status ReadEncryptedProto(Env* env, const string& fname,
      ::tensorflow::protobuf::MessageLite* proto) {
    string data;
    TF_RETURN_IF_ERROR(ReadFileToString(env, fname, &data));
    DecryptData(&data);

    if (!proto->ParseFromString(&data)) {
      TF_RETURN_IF_ERROR(stream->status());
      return errors::DataLoss("Can't parse ", fname,
      " as binary proto"); } return Status::OK();
    }
```

要使用这段代码，开发者需要自己定义 DecryptData()函数。它可以像下面的代码一样简单：

```
void DecryptData(string* data) {
     for (int i = 0; i < data.size(); ++i) {
     data[i] = data[i] ^ 0x23;
     }
}
```

上面的代码是一个概念性的展示，实际的代码会更复杂一些。

8.2.9　量化计算

神经网络中最有趣的研究领域之一就是如何降低模型精度。在默认情况下，用于计算的最方便的格式是 32 位浮点数，但是由于大多数网络在训练后对噪声具有容错性，事实证明很多推理可以在 8 位或更少的情况下运行，准确性没有太大的损失。我们在 11.2.1 节中谈到过这个问题，介绍了如何通过 quantize_weights 变换来缩小模型的文件大小。8 位缓冲区在用于计算之前会扩展到 32 位浮点数，因此对网络的变化是相当小的。所有其他的操作只是看到正常的浮点输入。

更激进的方法是，尝试使用 8 位表示法进行尽可能多的计算。许多 CPU 都有 SIMD 指令（如 NEON 或 AVX2），它们可以在每个周期执行更多的 8 位计算。这也意味着像 Qualcom 的 HVX DSP 或谷歌的张量处理单元这样的专门硬件可能不能很好地支持浮点运算，不能加速神经网络的计算。

从理论上讲，我们没有理由用少于 8 位的数据进行运算，实际上在很多实验中，我们已经看到 7 位甚至 5 位都是可用的，且没有太多的损失。然而，我们没有足够的硬件可以支持这些少于 8 位的运算单位。

1. 量化挑战

量化方法所面临的最大挑战是，神经网络中相当随机的数字范围，这些数字事先是不知道的，所以将它们拟合成 8 位非常困难。创建算术运算来使用这些表示也是非常棘手的，因为我们必须重新实现在使用浮点时免费获得的许多实用程序，例如范围检查。因此，优化后的代码与等效的浮动版本会有很大的不同。还有一个问题是，由于我们不知道什么时候会输入什么，所以中间计算的范围很难估计。

2. 量化表示

因为我们会处理数值较大的数组，这些数值通常分布在一个共同的范围内，所以可以使用浮点值的最小值和最大值将这些值线性地编码为 8 位。实际上，这看起来很像一个块浮点表示，尽管实际上它更灵活一些。下面是使用原始值对浮点数组进行编码的示例：

```
[-10.0, 20.0, 0.0]
```

读者看一下这个数组，可以看到最小值和最大值分别是−10.0 和 20.0。取数组中的每个值分别减去最小值−10，除以最大值和最小值之间的差值（这里是 20−(−10)=30），得到 0.0~1.0 之间的归一化值，然后分别乘以 255，转换成 8 位数值。算式如下：

```
[((-10.0 - -10.0) / 30.0) * 255, ((20.0 - -10.0) / 30.0) * 255,  ((0.0 - -10.0) / 30.0) * 255]
```

结果如下：

```
[0, 255, 85]
```

在处理这种量化表示时，最关键的一点是记住如果我们不知道它所基于的最小值和最大值，这种表示就没有意义。最好把这种方法想象成一个实数的压缩方案，只有在最大和最小值都定义的情况下，数值才有意义。

在 TensorFlow 中，这意味着每次量化张量通过图形时，都需要确保有两个辅助张量来存储最小和最大浮点数。所以所有接受张量作为输入的的缓冲器都要支持辅助张量。同时输出的定点值也有相应的辅助变量可以得到最大和最小值。

3. 数值代表性的不足

这种表示的好处是非常明显的。如果将最小值和最大值设置为 2 的幂，则可以使用它来保存传统的定点值，该范围不必像典型的有符号表示那样对称，并且可以适应几乎任何比例值。

当我们开始使用量化的时候，这些属性是非常重要的，因为我们没有清楚地理解我们可以把什么约束放在这些值上，而不会失去整体的精确性。随着我们获得更多的经验，我们意识到我们可以在不明显影响准确性的情况下进行更严格的限制，而且这种灵活的格式也有缺点。

在神经网络中，0 是一个特殊的数字，因为它被用于填充超出卷积矩阵的图像数值，并且是来自 Relu 激活函数的任何负数的输出。这就产生了一个微妙的问题，即如果 0 没有一个确切的表示，例如，如果最接近的编码值实际解码为 0.1 而不是 0.0，那么这个错误会损害网络整体的准确性。

量化在神经网络上运算时，由四舍五入引起的误差类似于他们在训练时产生的噪声。这意味着量化误差必须大致均匀，或者在足够大的运行中至少平均为零。如果数值编码范围内的每个数字出现的频率相同，并且当 0 比任何其他数字都多得多时，量化误差就可能被放大。这种结果会导致系统偏差，最终导致产生错误的结果。

另一个缺点是，可能会造成无意义的或无效的数值表示。例如，在最小值和最大值相等或最小值大于最大值的情况下，这种数值表示可能显示这种表示本身就是一种缺陷。

此外，当我们试图用相同的数值表示来表现 32 位数字时，数值表示的范围可能会变得非常大且很难实现。

为了解决这些问题，我们可以限制最小值和最大值可以达到的值。例如，当我们定义包括 0 的范围的数值范围时，如果最小值和最大值太靠近，那么可以将它们分开一点。将来，我们也可以强制数值对称分布。幸运的是，我们能够在不改变数值表示的情况下解决这个问题。在代码进行量化计算时，要尽可能强制这些规则，以获得更好的计算结果。

8.2.10 使用量化计算

生成定点模型的方法通常是，采用一个经过浮点数训练的模型，使用变换工具进行浮点到定点的转换。这个过程要尽可能用量化运算取代浮点数运算。由于 8 位算法的实现与浮点数完全不同，所以我们把转换的重点放在流行模型中常用的操作上，以下是最新的 8 位操作的列表：

- BiasAdd
- Concat
- Conv2D
- MatMul
- Relu
- Relu6
- AvgPool
- MaxPool
- Mul

这些操作足以实现 Inception 模型和许多其他卷积模型。在这类网络结构中，相邻 Ops 之间，我们使用 8 位量化的缓冲区来传递数值。如果遇到不支持的操作，任何量化的张量都将被转换为浮点数，并以正常的方式运行，在下一个 8 位操作之前再转换回量化计算。

我们来看一个 Relu 操作的例子。所有的 Relu 所做的是取一个张量数组并输出其输入的一个拷贝，但任何负数都被 0 代替。量化这个操作所做的第一件事是，用等效的 8 位版本（称为量化的 Relu）替换 Relu。此时，我们暂不关心和它相关的操作，它的输入和输出仍是浮点数。

我们使用量化操作来处理输入，这个操作输出浮点最小值和最大值，以及 8 位的编码值。然后由 QuantizedRelu 进行计算操作，它将 8 位值与范围一起输出。实际上，对于这个实现，输出范围与输入范围相同，所以我们可以直接将来自量化的最小值和最大值作为输入来量化，但是为了保持执行一致，对于每个量化输出总有一个约定的数值范围。

QuantizedRelu 编码的 8 位值与数值范围一起被送入 Dequantize 操作，产生最终的浮点输出，这种处理方式非常复杂。重要的是，这是一种通用的方法，我们先把浮点数量化并转换成定点数，进行定点数计算，然后把定点数结果转换成浮点数。我们可以只关注其中的 8 位的计算操作，而不必了解相关节点的运算。

通常，我们会遍历一遍 TensorFlow 的计算图，把运算做一次替换，然后通过结果图去除效率低下的操作。通过发现这种模式，转换工具可以删除那些不必要的操作，生成简化的计算图。

1. 去量化的重要性

消除量化计算非常重要，因为这意味着 8 位运算图的性能取决于模型中有多少操作具有量化的等效性，如果未转换的操作数在浮点和 8 位之间引起大量的转换，那么在运算图的开头或结尾处进行一些转换不会太重要。例如，SoftMax 通常是工作量很小的操作，并作为运算图的最后一步，所以它通常不是一个瓶颈。但是，如果再在运算图的核心部分混合使用浮点和 8 位进行计算，会对数值低位上的精确度造成影响，就失去了使用 8 位定点数的意义。

需要注意的一个问题是，要有效地运行 8 位算法需要系统特定的 SIMD 代码的支持。我们使用 gemmlowp 库来实现组成大量神经网络计算的矩阵乘法，但是目前只针对 ARM NEON 和移动 Intel 芯片进行了优化。实践中，我们使用 Eigen 针对这些芯片进行了高度优化的浮点库来提高计算质量。

2. 激活范围

上面还没有提到的一个挑战是，一些采用 8 位输入的操作实际上产生了 32 位输出。例如，如果开发者正在进行矩阵乘法，则每个输出值将是一系列相互相乘的 8 位输入数字的和。乘以 2 个 8 位输入的结果是一个 16 位的值，为了准确地累加它们中的一些需要大于 16 位的数据，这在大多数芯片上意味着要使用一个 32 位数值的计算单元。然而，使用这个结果作为输入的后续量化操作并不需要 32 位数据，因为对后续操作强制进行浮点数运算并不一定更有效，而且处理起来也相当困难。相反，我们通常将这 32 位转换为 8 位数进行运算。

可以想象计算一个特定矩阵乘法产生的最小和最大的可能值，并把这些值看成较大范围数值中的一个可以用 8 位数表示的数值。然而，这种编码是比较低效的，因为大多数神经网络操作的实际输入不具有极端的分布，所以我们实践中遇到的最小和最大的值将比他们的理论极限小得多。使用极端最大和最小值的范围意味着实际计算中的大部分数值将被浪费。

为了解决这个问题，我们需要知道通常遇到的数据极端情况。不幸的是，这些信息已经被证明是非常难以计算和分析的，所以我们最终不得不在实际的测试中检验和观察统计数据，我们只能通过运行整个神经网络来解决这个问题。常用的处理方法有以下三种。

（1）动态范围

确定动态值范围最简单的方法是，在生成一个 32 位操作之后，插入一个运算操作来计算出这些值的实际范围。然后可以将这个范围输入一个重新量化的操作中，该操作将 32 位张量转换为 8 位，并给定一个目标输出范围。

这种方法的一大优点是，不需要额外的数据或用户干预，所以这是 quantize_nodes 变换使用的默认方式。这种方法使得浮动网络变得简单，并将其转换为 8 位，然后可以检查精度和性能。

这种方法的缺点是，每次执行推断都需要运行范围计算，即查看每个输出值并计算每个缓冲区的最小值和最大值。这是额外的工作，在 CPU 上会引起性能降低，虽然这不是严重的性能损失，但是对于某些专用的机器学习的硬件平台来说，由于这些硬件可能无法处理这种动态的数值转换，所以会消耗额外的计算资源而引起严重的性能损失。。

（2）观察值范围

确定观察值范围最简单的方法是，通过网络运行一组具有代表性的数据，跟踪每个操作的范围随着时间的推移，使用统计方法来估计合理的值来覆盖所有这些范围，而不会浪费太多的精度。不幸的是，这种方法很难自动完成，因为构成代表性数据的思想依赖于应用程序。

例如，对于图像识别网络，以随机数或极限值作为输入的情况下，只会使用少数的模式识别路径，所以得到的范围将不是很有用。相反，需要使用具有代表性的数据，例如训练数据。

为了做到这一点，TensorFlow 提供了一个多阶段的方法，即使用 insert_logging 规则添加每次运行模型时都会输出范围值的调试操作。接下来我们演示一个完整的在预训练的 InceptionV3 上实现的例子。

首先，下载并解压缩模型文件：

```
mkdir /tmp/model/
curl "https://storage.googleapis.com/download.tensorflow.org/ \
     models/inception_dec_2015.zip" \
     -o /tmp/model/inception_dec_2015.zip
     unzip /tmp/model/inception_dec_2015.zip -d /tmp/model/
```

然后，使用 8 位计算进行量化：

```
bazel build tensorflow/tools/graph_transforms:transform_graph
bazel-bin/tensorflow/tools/graph_transforms/transform_graph \
    --in_graph="/tmp/model/tensorflow_inception_graph.pb" \
    --out_graph="/tmp/model/quantized_graph.pb" --inputs='Mul:0' \
    --outputs='softmax:0' --transforms='add_default_attributes
strip_unused_nodes(type=float, shape="1,299,299,3")
remove_nodes(op=Identity, op=CheckNumerics)
fold_old_batch_norms
quantize_weights
quantize_nodes
Strip_unused_nodes'
```

一旦完成，运行 label_image 示例以确保模型仍然提供预期的结果。它默认运行在 Grace Hopper 的图片上，所以 Uniform 是最好的输出结果：

```
bazel build tensorflow/examples/label_image:label_image
bazel-bin/tensorflow/examples/label_image/label_image \
    --input_mean=128 --input_std=128 --input_layer=Mul \
    --output_layer=softmax --graph=/tmp/model/quantized_graph.pb \
    --labels=/tmp/model/imagenet_comp_graph_label_strings.txt
```

接下来，我们把日志操作附加到所有 RequantizationRange 节点的输出上：

```
bazel-bin/tensorflow/tools/graph_transforms/transform_graph \
    --in_graph=/tmp/model/quantized_graph.pb \
    --out_graph=/tmp/model/logged_quantized_graph.pb \
    --inputs=Mul \
    --outputs=softmax \
    --transforms='insert_logging(op=RequantizationRange, \
    show_name=true, message="__requant_min_max:")'
```

现在可以运行该图了，stderr 应该包含日志记录，显示该范围在该运行时的值。

```
bazel-bin/tensorflow/examples/label_image/label_image \
```

```
    --input_mean=128 --input_std=128 \
    --input_layer=Mul --output_layer=softmax \
    --graph=/tmp/model/logged_quantized_graph.pb \
    --labels=/tmp/model/imagenet_comp_graph_label_strings.txt 2> \
    /tmp/model/logged_ranges.txt
cat /tmp/model/logged_ranges.txt
```

我们会看到一系列这样的行输出：

```
;conv/Conv2D/eightbit/requant_range__print*; *requant_min_max:
[-20.887871] [22.274715]
```

输出中，我们可以得到这些操作数值的范围值。这里只是在单个图像上运行一次计算图形。在实际应用中，我们希望运行数百个有代表性的图像，以获得所有常用数值的范围。

最后，使用 freeze_requantization_ranges 变换来获取我们收集的信息，并用简单的常量代替动态范围计算，代码如下：

```
bazel-bin/tensorflow/tools/graph_transforms/transform_graph \
    --in_graph=/tmp/model/quantized_graph.pb \
    --out_graph=/tmp/model/ranged_quantized_graph.pb \
    --inputs=Mul \
    --outputs=softmax \
    --transforms='freeze_requantization_ranges \
    (min_max_log_file=/tmp/model/logged_ranges.txt)'
```

如果在这个新的网络上运行 label_image 例子，应该能够看到一致的结果，尽管确切的计算数字可能与以前有些不同：

```
bazel-bin/tensorflow/examples/label_image/label_image \
    --input_mean=128 --input_std=128 --input_layer=Mul \
    --output_layer=softmax \
    --graph=/tmp/model/ranged_quantized_graph.pb \
    --labels=/tmp/model/imagenet_comp_graph_label_strings.txt
```

（3）训练值范围

关于训练值范围，我们以计算激活层数值的方法来举例。

计算激活层数值范围的方法是，在训练过程中发现数值范围。开发者可以使用 FakeQuantWithMinMaxVars 的 Ops 来执行此操作。这个 Ops 操作可以放在图中的不同点上，在由两个代表最小值和最大值的变量输入设置的范围内，通过将浮点数的输入值转换成一个定点数的方法来模拟量化不准确性。这些输入值范围会在更新的过程中根据设定的最大值、最小值而变化。

不幸的是，在 Python 中没有任何方便的函数可以将这些操作添加到你的模型中，所以如果读者采用这种方法，就要在特定的范围内手动插入操作和变量。

这种方法有两个好处，可以弥补它在使用中的不便。当我们采用预训练的浮点数模型并简单地将其转换为 8 位时，精度通常会有小幅度的下降，例如，InceptionV3 上的 top-1 精度从 78%下降到 77%。但如果把量化集成进训练通常会大大缩小损失，或者事先知道数值范围也有助于减少推断过程中的延迟。

8.3　设计机器学习应用程序要点

在设计机器学习的应用时，我们要注意以下两点：

（1）如果必须使用人工智能，我们的目标是什么？我们必须要问自己这个问题。请注意以上我们讨论过的这些问题，并且开发者可能还要处理许多移动设备的特定问题。因此请确定机器学习是正确的且是唯一的解决方案。

基于规则的解决方案可以解决一些问题。如果开发者一定要在问题上应用人工智能，可以先尝试一个简单的解决方案或进行更多离线测试，再开始在移动设备上工作。

开发者要选择正确的机器学习模型和方式。比如，开发者只需要进行推理，或者要同时进行推理和训练。选择正确且恰当的机器学习方式，会极大减轻开发的难度。

（2）尝试尽可能减少应用的大小。大型应用通常意味着更大的内存需求，响应慢、低性能和糟糕的用户体验。我们需要确定应用增加的部分是由应用程序或机器学习引起的，通常机器学习可以因为引入大模型的文件而增大。我们可以通过设置“SELECTIVE_REGISTRATION”删除模型文件中的不必要的操作。

如果开发者在应用中嵌入机器学习模型，用户将首先体验到慢速下载。解决方法是，用户在第一次运行应用程序时下载模型。在这种情况下，如果网络连接不稳定或足够快，开发者还需要准备一个后备计划。后备计划可以是基于规则的解决方案，也可以是非常小的机器学习解决方案。

在 Android 上一般使用 Java API。通过使用原生 C/C++ API，开发者可能会增加应用大小，并且开发者必须为不同的平台/硬件构建原生代码。在测试中，我们发现调用 Java API 不一定会使应用程序变慢，性能的瓶颈通常是运行模型。

同时，我们还要考虑一些其他因素：

- 重复使用内存碎片。
- 模型可以存储在资源文件夹或临时文件夹中，Java 有 API 支持加载模型文件，用户不必使用原始 API 加载模型。
- 必要时使用原生 API OpenGL。
- 通过硬件加速或运算量化来提高性能。
- 尊重 Android 运行规则，尽量节省电池。
- 保护模型。通常，机器学习执行机制不知道运行图中的上下文，只需遵循应用程序给出的任何命令。如果模型被修改或遭黑客攻击，则输出结果肯定不是我们所期望的。如果使用下载模型，则可能需要在执行模型之前进行健全性检查。
- 防止模型滥用。如果模型和应用是分开的，开发者可能希望保护它免受滥用。而当模型在相对安全的数据中心运行时，开发者可能不会遇到这种挑战。

第 9 章 TensorFlow 的硬件加速

本章介绍如何利用现有的移动设备构建和运行一个机器学习模型，以及如何使用硬件加速。现在主流的移动设备中，有两种设备支持硬件加速，华为和高通 Qualcomm。下面介绍一下如何在这两种设备上开发和运行机器学习的应用。

9.1 神经网络接口

Android 神经网络接口（Android Neural Networks API，简称 NNAPI）是一个 Android C API，专门为在移动设备上运行机器学习的密集型运算而设计的 API。NNAPI 旨在为构建和训练神经网络的更高级机器学习框架（例如 TensorFlow Lite、CAFFE2 或其他）提供一个基础的功能层。API 适用于运行 Android 8.1（API 级别为 27）或更高版本的所有设备。

NNAPI 推理（Inference）的过程是，使用开发者已训练的自定义模型，从 Android 设备中读取数据、运行模型并得到结果。一些典型的推理的应用例子包括图像分类、预测用户行为，以及选择对搜索查询的适当响应等。

在移动设备上推理具有如下优势：

- 延迟时间：不需要通过网络连接发送请求并等待响应。这对处理从摄像头传入的连续帧的视频应用至关重要。
- 可用性：应用甚至可以在没有网络覆盖的条件下运行。
- 速度：与单纯的通用 CPU 相比，特定于神经网络处理的新硬件可以提供显著加快的计算速度。
- 隐私：数据不会离开设备。
- 费用：所有计算都在设备上执行，不需要额外服务器。

但是，开发者也应考虑它一些的利弊：

- 系统利用率：神经网络包含很多计算，这会增加电池消耗。开发者的应用需要注意耗电量，应当考虑监视电池的运行状况，尤其要针对长时间运行的计算进行监视。
- 应用大小：注意模型的大小。模型可能会占用很多空间。由于在开发者的 APK 中绑定较大的模型会过度地影响用户，所以开发者需要考虑在应用安装后下载模型、使用较小的模型，或在云中运行计算。但 NNAPI 未提供在云中运行模型的功能。

9.1.1 了解 Neural Networks API 运行时

NNAPI 的目的是被机器学习代码、框架和工具调用，这些可以让开发者脱离物理设备，而训练他们的模型并将模型部署在 Android 设备上。Android 的应用一般不会直接使用 NNAPI，但会使用更高级的机器学习框架。这些框架反过来可以使用 NNAPI 在被支持的设备上执行硬件加速的推理运算。

根据应用的要求和设备上的硬件能力，Android 的神经网络运行时（Runtime）可以在可用的处理器（包括专用的神经网络硬件、图形处理单元（GPU）和数字信号处理器（DSP）

之间有效地分配计算工作负载。对于缺少专用的供应商驱动程序的设备，NNAPI 运行时（Runtime）将会在 CPU 上运行优化后的代码。NNAPI 的系统架构如图 9-1 所示。

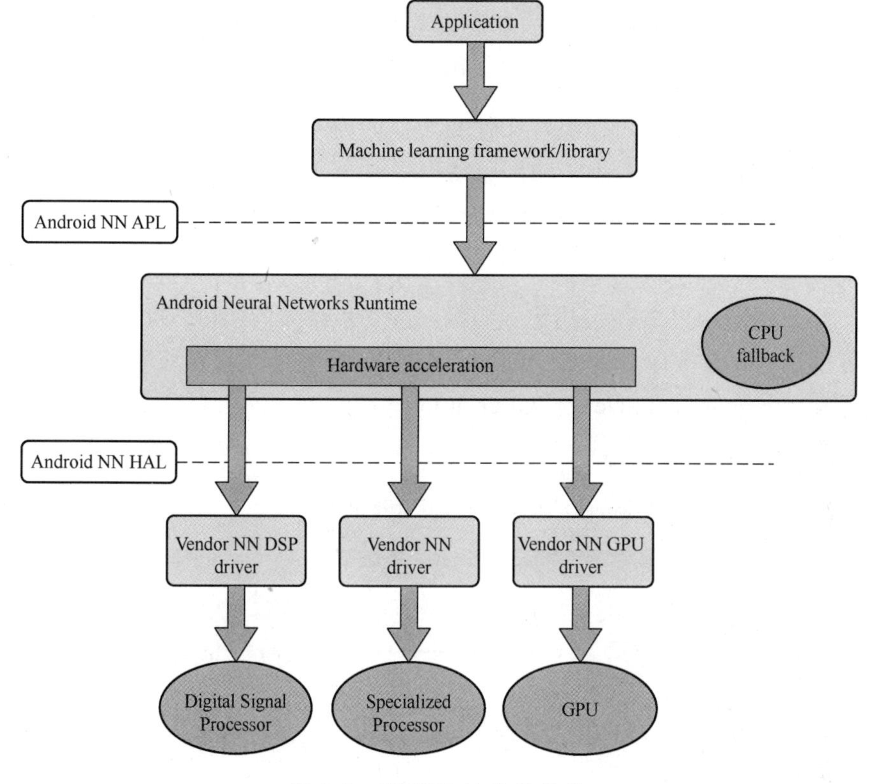

图 9-1　NNAPI 的系统架构

9.1.2　Neural Networks API 编程模型

要使用 NNAPI 执行计算，开发者首先需要构建一个可以定义要执行的计算的有向图（Directed Graph）。此计算图与输入数据（比如从机器学习框架传递过来的权重和偏差）相结合，构成 NNAPI 运行时评估的模型。

NNAPI 定义了如下四个概念：

模型：是数学运算和通过训练过程学习的常量值的计算图。这些运算特定于神经网络。它们包括二维（2D）卷积、逻辑（Sigmoid）激活和全连接（ReLU）激活等。创建模型是一个同步操作，一旦创建成功就可以在线程和编译之间重用模型。在 NNAPI 中，一个模型表示为一个 ANeuralNetworksModel 实例。

编译：表示用于将 NNAPI 模型编译到更低级别的代码中。创建编译是一个同步操作，一旦成功创建就可以在线程和执行之间重用编译。在 NNAPI 中，每个编译表示为一个 ANeuralNetworksCompilation 实例。

内存：表示共享内存、内存映射文件和类似的内存缓冲区。使用内存缓冲区可以让 NNAPI 在运行时将数据更高效地传输到驱动程序中。一个应用通常会创建一个共享内存缓冲区，其中包含定义模型所需的每一个张量。也可以使用内存缓冲区来存储执行实例的输入和输出。在 NNAPI 中，每个内存缓冲区表示为一个 ANeuralNetworksMemory 实例。

执行：用于将 NNAPI 模型应用到一组输入并采集结果的接口。执行是一种异步操作。多个线程可以在相同的执行上等待。当执行完成时，所有的线程都将释放。在 NNAPI 中，每一个执行表示一个 ANeuralNetworksExecution 实例。

基本的编程流程如图 9-2 所示。

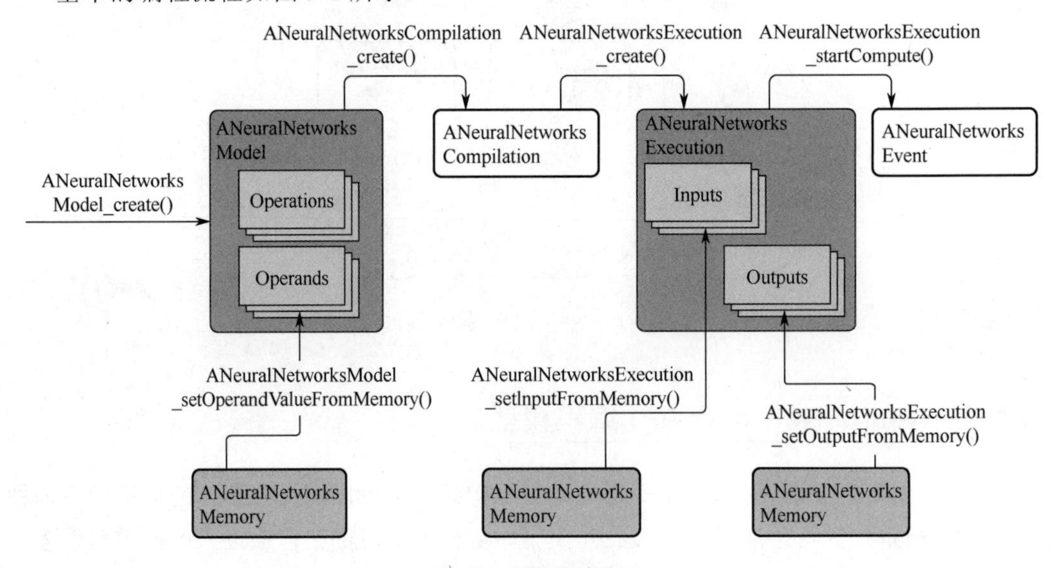

图 9-2 编程流程

下面的代码是 NNAPI 接口的定义。请注意，现在的版本是 1.2 版本。其中的代码注释被删除了，只保留代码的核心部分。

```
package android.hardware.neuralnetworks@1.2;

import @1.0::ErrorStatus;
```

```
import @1.0::IPreparedModelCallback;
import @1.1::ExecutionPreference;
import @1.1::IDevice;

/**
 * This interface represents a device driver.
 */
interface IDevice extends @1.1::IDevice {
   getVersionString() generates (ErrorStatus status, string version);


   getSupportedOperations_1_2(Model model)
         generates (ErrorStatus status, vec<bool> supportedOperations);

   prepareModel_1_2(Model model, ExecutionPreference preference,
               IPreparedModelCallback callback)
        generates (ErrorStatus status);
};
```

9.1.3　NNAPI 实现的实例

本节介绍一个简单的 NNAPI 实现的实例。

1. 下载和安装实例

首先，下载 android-ndk.git，nn_sample 文件夹里有简单的 NNAPI 的实现；然后，使用 Gradle（命令为 gradlew）构建并生成 app/build/outputs/apk/debug/app-debug.apk。具体代码如下：

```
$ git clone https://github.com/googlesamples/android-ndk.git
$ cd android-ndk/nn_sample
$ ./gradlew build
```

由于这个例子需要编译原生 C 代码，例子里使用了 cmake，所以要安装 cmake 和 ninja。在 Ubuntu 上可以通过执行命令“sudo apt-get cmake”来安装 cmake，安装 ninja 可以到 https://github.com/ninja-build/ninja/releases 下载。

除了使用 Gradle，我们还尝试使用 Bazel 去编译这个应用，看看有什么不同。使用 Bazel 之前，要新建 workspace 和 build 文件，TensorFlow Lite 已经把这些都做好了，我们可以借用一下。把 nn_sample 文件夹拷贝到 tensorflow/examples 下，在 tensorflow/examples/nn_sample/app/src/main 下新建一个 build 文件，把下面的内容复制到 build 文件中：

```
package(default_visibility = ["//visibility:public"])

licenses(["notice"])  # Apache 2.0

load(
    "//tensorflow:tensorflow.bzl",
    "tf_copts",
    "tf_opts_nortti_if_android",
)

exports_files(["LICENSE"])

LINKER_SCRIPT = "//tensorflow/contrib/android:jni/version_script.lds"

cc_binary(
    name = "libnn_sample.so",
    srcs = glob([
        "cpp/**/*.cpp",
        "cpp/**/*.h",
    ]),
    copts = tf_copts(),
    defines = ["STANDALONE_DEMO_LIB"],
    linkopts = [
        "-landroid",
        "-ldl",
        "-ljnigraphics",
        "-llog",
        "-lm",
        "-lneuralnetworks",
        "-z defs",
        "-s",
        "-Wl,--version-script",  # This line must be directly followed by
LINKER_SCRIPT.
        "$(location {})".format(LINKER_SCRIPT),
    ],
    linkshared = 1,
    linkstatic = 1,
```

```
    tags = [
        "manual",
        "notap",
    ],
    deps = [
        LINKER_SCRIPT,
    ],
)

cc_library(
    name = "libnn_sample",
    srcs = [
        "libnn_sample.so",
    ],
    tags = [
        "manual",
        "notap",
    ],
)

android_binary(
    name = "nnapi_demo",
    srcs = glob([
        "java/**/*.java",
    ]),
    aapt_version = "aapt",
    assets = [
        ":model_assets",
    ],
    assets_dir = "assets",
    custom_package = "com.example.android.nnapidemo",
    inline_constants = 1,
    manifest = "AndroidManifest.xml",
    nocompress_extensions = [
        ".bin",
    ],
    resource_files = glob(["res/**"]),
    tags = [
        "manual",
        "notap",
```

```
        ],
        deps = [
            ":libnn_sample",
            "@androidsdk//com.android.support.constraint:constraint-
layout-1.0.2",
        ],
    )

    filegroup(
        name = "model_assets",
        srcs = [
            "assets/model_data.bin",
        ],
    )
```

Gradle 工具的使用依赖于'com.android.support.constraint:constraint-layout:1.1.3'，原生的 Android NDK 不自带这个版本，但是由于不影响计算结果，所以这里直接使用“@androidsdk//com.android.support.constraint:constraint-layout-1.0.2”来替代 1.1.3 版本，读者也可以从 maven 下载使用最新版本。

这里还要指定 nocompress_extensions 参数禁止压缩带有“.bin”的文件。使用 cc_binary 构建二进制文件 libnn_sample.so，使用 cc_library 构建动态链接库。现在，执行下面的命令：

```
    $ bazel build --fat_apk_cpu=arm64-v8a --android_cpu=arm64-v8a
//tensorflow/examples/nn_sample/app/src/main:nnapi_demo
```

生成文件 nnapi_demo.apk，使用命令“unzip –l”去查看文件内容，APK 的 assets 里应该包含 model_data.bin。读者可以将 nnapi_demo.apk 文件和 Gradle 构建的文件进行比较：

```
    $ unzip -l bazel-bin/tensorflow/examples/nn_sample/app/src/main/
nnapi_demo.apk
      Length      Date    Time    Name
    --------- ---------- -----   ----
       194720  2010-01-01 00:00   lib/arm64-v8a/libnn_sample.so

    $ unzip -l ./app/build/outputs/apk/release/app-release-unsigned.apk
      Length      Date    Time    Name
    --------- ---------- -----   ----
       223312  1980-00-00 00:00   lib/arm64-v8a/libnn_sample.so
```

Bazel 构建的文件大小少于 200KB，Gradle 构建的大于 200KB。同样的技术也可以用于 Gradle 去减少 APK 的文件大小，只是需要改进 Gradle 的构建过程，而这些 TensorFlow 里的 build 已经都做好了。

2. 模型准备

这个实例演示了一个简单的 NNAPI 的实现。程序入口是 nn_sample/app/src/main/java/com/example/android/nnapidemo/MainActivity.java。在 onCreate 里加载模型文件 model_data.bin，代码如下：

```
    extern "C"
    JNIEXPORT jlong
    JNICALL
    Java_com_example_android_nnapidemo_MainActivity_initModel(
            JNIEnv *env,
            jobject /* this */,
            jobject _assetManager,
            jstring _assetName) {

    // 获取模型数据文件的文件描述符
        AAssetManager     *assetManager    =     AAssetManager_fromJava(env,
_assetManager);
        const char *assetName = env->GetStringUTFChars(_assetName, NULL);
        AAsset    *asset    =    AAssetManager_open(assetManager,    assetName,
AASSET_MODE_BUFFER);
        if(asset == nullptr) {
            __android_log_print(ANDROID_LOG_ERROR, LOG_TAG, "Failed to open the
asset.");
            return 0;
        }
        env->ReleaseStringUTFChars(_assetName, assetName);
        off_t offset, length;
        int fd = AAsset_openFileDescriptor(asset, &offset, &length);
        AAsset_close(asset);
        if (fd < 0) {
            __android_log_print(ANDROID_LOG_ERROR, LOG_TAG,
                                "Failed to open the model_data file descriptor.");
            return 0;
        }
```

```
    SimpleModel* nn_model = new SimpleModel(length, PROT_READ, fd, offset);
    if (!nn_model->CreateCompiledModel()) {
        __android_log_print(ANDROID_LOG_ERROR, LOG_TAG,
                            "Failed to prepare the model.");
        return 0;
    }

    return (jlong)(uintptr_t)nn_model;
}
```

从上面的代码中可以看到，我们使用 Android NDK 的 API AAssetManager_open 和 AAsset_openFileDescriptor 读取模型文件，构建一个 SimpleModel 类。此应用先调用 Java_com_example_android_nnapidemo_MainActivity_startCompute，再调用 SimpleModel 类的 Compute 去实现 tensor 的运算。

这个应用的实现非常简单，就是(tensor0 + tensor1) × (tensor2 + tensor3)，其中 tensor0 和 tensor2 是常量，是从模型文件 model_data.bin 中读入的，它们在实际的应用中被看作训练后的模型权重（Weights）。

```
     (tensor0 + tensor1) * (tensor2 + tensor3)的图形表示如下:
 tensor0 ---+
            +--- ADD ---> intermediateOutput0 ---+
 tensor1 ---+                                    |
                                                 +--- MUL---> output
 tensor2 ---+                                    |
            +--- ADD ---> intermediateOutput1 ---+
 tensor3 ---+
```

上面这个图形表示可以看作一个小模型，在这个模型里一共有 8 个算子，分别为：

- 2 个张量，分别供给两个加法。
- 2 个常量张量， 分别供给两个加法。
- 1 个混合激活算子，被用于加法和乘法。
- 2 个中间结果张量，分别代表了两个加法的结果。这两个张量也供给乘法。
- 1 个乘法的结果。

TensorFlow 的两个重要的概念就是模型和张量，SimpleModel 类也是围绕这两个概念进行设计的。下面是 bool SimpleModel::CreateCompiledModel()函数的实现代码：

```
    // 创建 AneralNetworksModel 句柄
     status = ANeuralNetworksModel_create(&model_);
     if (status != ANEURALNETWORKS_NO_ERROR) {
       __android_log_print(ANDROID_LOG_ERROR,                          LOG_TAG,
"ANeuralNetworksModel_create failed");
       return false;
     }
```

上面代码生成一个空的模型，然后，向这个模型里填装算子：

```
        // 为张量添加操作数
        status = ANeuralNetworksModel_addOperand(model_, &float32TensorType);
        uint32_t tensor0 = opIdx++;
        if (status != ANEURALNETWORKS_NO_ERROR) {
            __android_log_print(ANDROID_LOG_ERROR, LOG_TAG,
                            "ANeuralNetworksModel_addOperand    failed    for
operand (%d)",
                            tensor0);
            return false;
        }
```

上面代码实现了在模型里生成运算数。在生成了所有必要的运算数之后，接下来生成运算符：

```
       // 添加第一个 ADD operation
        std::vector<uint32_t> add1InputOperands = {
                tensor0,
                tensor1,
                fusedActivationFuncNone,
        };
        status  =  ANeuralNetworksModel_addOperation(model_,  ANEURALNETWORKS_
ADD,
                                          add1InputOperands.size(),
add1InputOperands.data(),
                                          1, &intermediateOutput0);
        if (status != ANEURALNETWORKS_NO_ERROR) {
            __android_log_print(ANDROID_LOG_ERROR, LOG_TAG,
                            "ANeuralNetworksModel_addOperation   failed   for
ADD_1");
            return false;
        }
```

上面代码向模型里增加了一个加法，加法的输入运算数一个是 tensor0，另一个是 tensor1，加法的结果是 intermediateOutput0。用这种方法，代码把所有的运算符和关联的运算数都填充进模型，然后告诉模型输入和输出的张量。具体实现代码如下：

```
    // 识别模型的输入和输出张量
        // Inputs: {tensor1, tensor3}
        // Outputs: {multiplierOutput}
        std::vector<uint32_t> modelInputOperands = {
                tensor1, tensor3,
        };
        status = ANeuralNetworksModel_identifyInputsAndOutputs(model_,

modelInputOperands.size(),

modelInputOperands.data(),
                                                              1,
                                                              &multiplierOutput);
        if (status != ANEURALNETWORKS_NO_ERROR) {
            __android_log_print(ANDROID_LOG_ERROR, LOG_TAG,
                                "ANeuralNetworksModel_identifyInputsAndOutputs
failed");
            return false;
        }
```

从上面代码可以看到，使用者需输入 tensor1 和 tensor3 的数值，tensor0 和 tensor2 是常量，multiplierOutput 是模型的输出。然后，把已生成的 model 转换成 ANeuralNetworksCompilation，代码如下：

```
    // 创建 AneralNetworksCompetition 对象
    status = ANeuralNetworksCompilation_create(model_, &compilation_);
        if (status != ANEURALNETWORKS_NO_ERROR) {
            __android_log_print(ANDROID_LOG_ERROR, LOG_TAG,
                                "ANeuralNetworksCompilation_create failed");
            return false;
        }

        // 编译完成
        status = ANeuralNetworksCompilation_finish(compilation_);
        if (status != ANEURALNETWORKS_NO_ERROR) {
            __android_log_print(ANDROID_LOG_ERROR, LOG_TAG,
                                "ANeuralNetworksCompilation_finish failed");
            return false;
        }
```

至此，模型的准备基本完成了。

3. 执行模型

首先，生成一个执行器，代码如下：

```
    // 从已编译的模型中创建 AneralNetworksExecution 对象
        // Note:
        //   1. 所有输入和输出数据都绑定到 AneralNetworksExecution 对象
        //   2. 可以从同一编译模型创建多个并发执行实例
        // 此示例只使用已编译模型的一次执行
        ANeuralNetworksExecution *execution;
        int32_t   status   =   ANeuralNetworksExecution_create(compilation_,
&execution);
        if (status != ANEURALNETWORKS_NO_ERROR) {
           __android_log_print(ANDROID_LOG_ERROR, LOG_TAG,
                                "ANeuralNetworksExecution_create failed");
           return false;
        }
```

然后，设定输入和输出，下面是示例代码。注意，这里尽量使用共享内存以节省数据拷贝。

```
    // AneralNetworksExecution_setinputfrommemory 将操作数与共享内存关联最小化原始
数据副本数的区域。注意：这里的索引“1”表示 modelinput 列表的第二个操作数
        status  =  ANeuralNetworksExecution_setInputFromMemory(execution,  1,
nullptr,
                                                   memoryInput2_, 0,
                                                   tensorSize_         *
sizeof(float));
        if (status != ANEURALNETWORKS_NO_ERROR) {
           __android_log_print(ANDROID_LOG_ERROR, LOG_TAG,
                                "ANeuralNetworksExecution_setInputFromMemory
failed for input2");
           return false;
        }
```

接下来，执行 ANeuralNetworksExecution_startCompute，代码如下：

```
    // 开始执行模型
    // 注意：这里的执行是异步的，并且将创建 AneralNetworksEvent 对象以监视执行状态
    ANeuralNetworksEvent *event = nullptr;
    status = ANeuralNetworksExecution_startCompute(execution, &event);
    if (status != ANEURALNETWORKS_NO_ERROR) {
```

```
    __android_log_print(ANDROID_LOG_ERROR, LOG_TAG,
                        "ANeuralNetworksExecution_startCompute failed");
    return false;
}

// 等待执行完成，实现同步调用
status = ANeuralNetworksEvent_wait(event);
if (status != ANEURALNETWORKS_NO_ERROR) {
    __android_log_print(ANDROID_LOG_ERROR, LOG_TAG,
                        "ANeuralNetworksEvent_wait failed");
    return false;
}
```

由于我们要调用 ANeuralNetworksEvent_wait 去等待运算的结束，所以运算要运行在不同的线程上。

最后，读取共享内存的数据：

```
float *outputTensorPtr = reinterpret_cast<float *>(mmap(nullptr,
                                             tensorSize_ * sizeof(float),
                                             PROT_READ, MAP_SHARED,
                                             outputTensorFd_, 0));
```

另外，不要忘记清理使用过的内存：

```
    ANeuralNetworksEvent_free(event);
    ANeuralNetworksExecution_free(execution);

    ANeuralNetworksCompilation_free(compilation_);
    ANeuralNetworksModel_free(model_);
    ANeuralNetworksMemory_free(memoryModel_);
    ANeuralNetworksMemory_free(memoryInput2_);
    ANeuralNetworksMemory_free(memoryOutput_);
```

9.2 硬件加速

本章主要讨论与硬件加速有关的进展和知识，其中将简要涉及高通（Qualcomm）、华为（Huawei NPU）及 ONNX 在硬件加速上的支持和相关的技术。

很多公司都在积极开发机器学习的硬件支持，有两家是公认的比较有影响力的。一个是华为 NPU，另一个是 Qualcomm DSP/GPU。

华为自主研发的人工智能芯片应用在华为手机上，取得了不错的市场效果，目前在国内处于领先地位。高通在手机芯片研发上有很长的历史，有很多技术积累。高通使用已有的 DSP 和 GPU 技术，也实现了不错的机器学习硬件加速功能。

在笔者看来，两家公司都独自研发了对硬件加速的支持，试图以自己的方式提供"AI"能力并尝试构建他们的生态系统，虽然具有商业意义，但对开发人员来说却很麻烦，这意味着开发人员必须在不同的设备上开发不同的应用程序。

笔者在华为设备上做了一些初步的尝试，本书记录了一些实验过程和结果。实验需要一个支持 NPU 的设备，笔者的设备是华为 Mate 10，通过执行命令"$ adb shell getprop | grep finger"，得到如下返回结果：

```
$  [ro.vendor.build.fingerprint]:[HUAWEI/ALP-AL00/HWALP:8.0.1/HUAWEIALP-AL00/.../release-keys]
```

9.2.1　高通网络处理器

市场上有很多手机都采用高通芯片组，找到使用 Qualcomm 芯片组的手机并不难，但是还需要找到支持 GPU 或 DSP 的设备，笔者可以参考下面的设备列表：

```
| Snapdragon  Device       | CPU     | GPU     | DSP           |
| -------------------------|:-------:|:-------:|:-------------:|
| Qualcomm Snapdragon 845  | Yes     | Yes     | Yes (CDSP)    |
| Qualcomm Snapdragon 821  | Yes     | Yes     | Yes (ADSP)    |
| Qualcomm Snapdragon 820  | Yes     | Yes     | Yes (ADSP)    |
| Qualcomm Snapdragon 710  | Yes     | Yes     | Yes (CDSP)    |
| Qualcomm Snapdragon 660  | Yes     | Yes     | Yes (CDSP)    |
| Qualcomm Snapdragon 652  | Yes     | Yes     | No          |
| Qualcomm Snapdragon 630  | Yes     | Yes     | No          |
| Qualcomm Snapdragon 636  | Yes     | Yes     | No          |
| Qualcomm Snapdragon 625  | Yes     | Yes     | No          |
| Qualcomm Snapdragon 450  | Yes     | Yes     | No          |
```

高通 Qualcomm 提供 SDK、示例代码等给开发人员。这里我们会介绍高通神经网络处理器（Snapdragon Neural Processing Engine，简称 SNPE）。开发者需要登录高通开发者网页（https://developer.qualcomm.com）去获取 SDK 等资源。高通登录页面如图 9-3 所示，登录以后就可以下载 SDK，查阅相关资源。

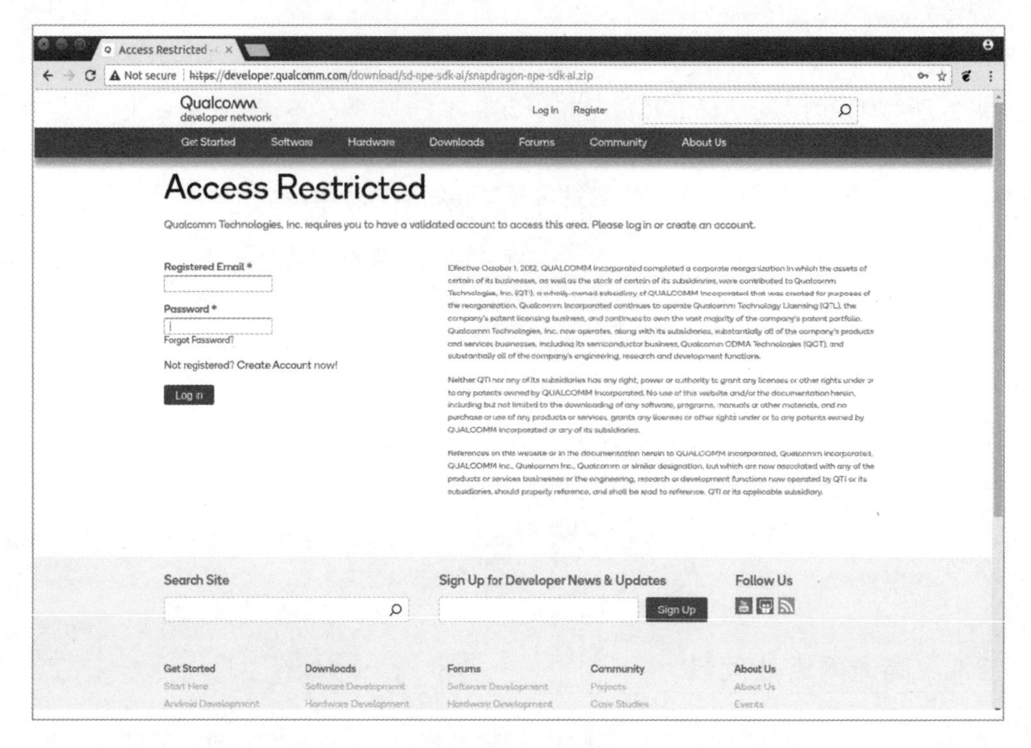

图 9-3 高通登录页面

Snapdragon neural Processing Engine（SNPE）的架构如图 9-4 所示。

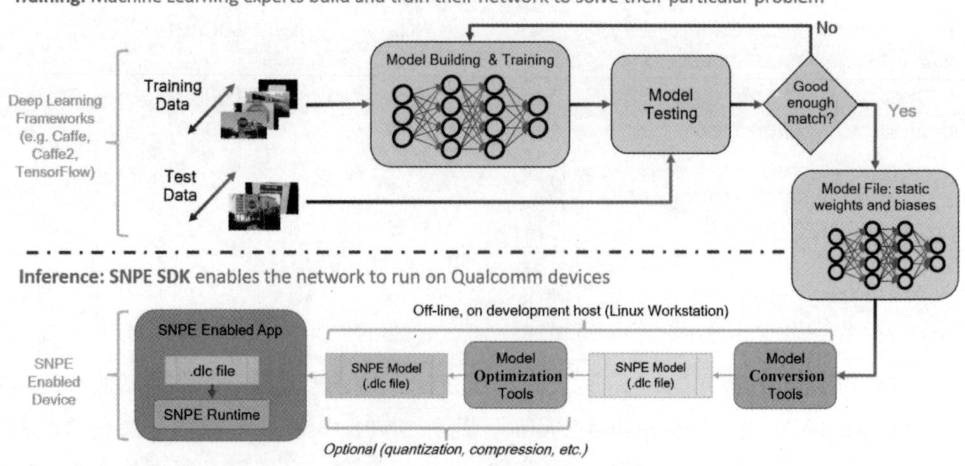

图 9-4 SNPE 架构图

SNPE 支持 TensorFlow、CAFFE 和 CAFFE2，SNPE 也支持 ONNX。SNPE 会将 CAFFE、CAFFE2、ONNXTM 和 TensorFlow 模型转换为 SNPE 深度学习容器（DLC）文件。读者最好同时安装 CAFFE 和 CAFFE2 两个框架，因为模型是基于 CAFFE2 的，并且 Android 的示例代码需要生成 alexnet dlc 文件，该文件将从一个脚本生成，而这个脚本需要 CAFFE。

笔者尝试过市场上的几款手机，笔者的建议是试试一款不是特别深度定制的手机，这会减少很多开发麻烦。最后笔者选择在小米的手机上做一些测试。测试 SNPE 之前，先运行命令“$ snpe-net-run –h”，看到正确的运行结果说明成功，如果没有输出任何东西，说明失败。如果读者的设备支持 DSP 或 GPU，现在就可以尝试使用它们。

首先，从开发者网页下载 Qualcomm Neural Processing SDK，将 snpe-1.25.1.zip 解压后，得到如下文件：

```
-rw-r--r--@  1  29911 May 31  2018 LICENSE.txt
-rwxr-xr-x@  1  47090 Apr 17 03:25 NOTICE.txt
-rwxr-xr-x@  1   3869 Jun  4 10:47 REDIST.txt
-rw-r--r--@  1   7018 Jun 11 18:19 ReleaseNotes.txt
drwxr-xr-x@  4    128 Jun 30 12:11 android
drwxr-xr-x@  7    224 Jun 30 12:11 benchmarks
drwxr-xr-x@ 12    384 Jun 30 12:11 bin
drwxr-xr-x@  3     96 Jun 30 12:11 doc
drwxr-xr-x@  5    160 Jun 30 12:11 examples
drwxr-xr-x@  3     96 Jun 30 12:11 include
drwxr-xr-x@ 11    352 Jun 30 12:11 lib
drwxr-xr-x@  5    160 Jun 30 12:11 models
drwxr-xr-x@  3     96 Jun 30 12:11 share
```

现在，我们可以运行 Android 示例应用程序。这个实例程序在 examples/android/image-classifiers 下，文件内容如下所示：

```
drwxr-xr-x@  6   192 Jun 30 12:11 app
-rw-r--r--@  1   412 Apr 17 03:25 build.gradle
drwxr-xr-x@  3    96 Jun 30 12:11 gradle
-rw-r--r--@  1   182 Apr 17 03:25 gradle.properties
-rwxr-xr-x@  1  4971 May 31  2018 gradlew
-rw-r--r--@  1  2404 May 31  2018 gradlew.bat
-rw-r--r--@  1   158 May 31  2018 settings.gradle
-rw-r--r--@  1   735 Apr 17 03:25 setup_alexnet.sh
-rw-r--r--@  1   894 May  7 08:24 setup_inceptionv3.sh
```

读者可以使用 Gradle 编译运行程序。通常，笔者会添加 Bazel 构建文件，所以读者也可以通过 Bazel 构建，方法如下：

首先，新建两个 build 文件。

一个是 app/src/main/下的 build 文件，内容如下：

```
ndroid_library(
    name = "demo_lib",
    srcs = glob([
        "java/**/*.java",
    ]),
    custom_package = "com.qualcomm.qti.snpe.imageclassifiers",
    manifest = "AndroidManifest.xml",
    resource_files = glob(["res/**"]),
    deps = [
        "//app/libs:snpe_release_aar",
        "//third_party:android_arch_core_common",
        "//third_party:android_arch_lifecycle_common",
        "//third_party:com_android_support_constraint_constraint_layout_
solver",
        "//third_party:com_android_support_support_annotations",
        "@android_arch_lifecycle_runtime//aar",
        "@com_android_support_appcompat_v7//aar",
        "@com_android_support_cardview//aar",
        "@com_android_support_constraint_constraint_layout//aar",
        "@com_android_support_design//aar",
        "@com_android_support_recyclerview//aar",
        "@com_android_support_support_compat//aar",
        "@com_android_support_support_core_ui//aar",
        "@com_android_support_support_core_utils//aar",
        "@com_android_support_support_fragment//aar",
        "@com_android_support_support_v4//aar",
        "@com_android_support_support_vector_drawable//aar",
    ],
)

android_binary(
    name = "demo",
    custom_package = "com.qualcomm.qti.snpe.imageclassifiers",
    manifest = "AndroidManifest.xml",
    manifest_values = {
        "applicationId": "com.qualcomm.qti.snpe.imageclassifiers",
    },
    multidex = "native",
    resource_files = glob(["res/**"]),
    deps = [
```

```
        ":demo_lib",
    ],
)
```

另外一个是 app/libs/下的 build 文件，文件内容如下：

```
package(default_visibility = ["//visibility:public"])

aar_import(
    name = "snpe_release_aar",
    aar = "snpe-release.aar",
)
```

编译执行文件：

```
$ bazel build app/src/main:demo
```

执行上面的命令后得到 Android 应用程序，安装 Android 应用，结果如图 9-5 所示。

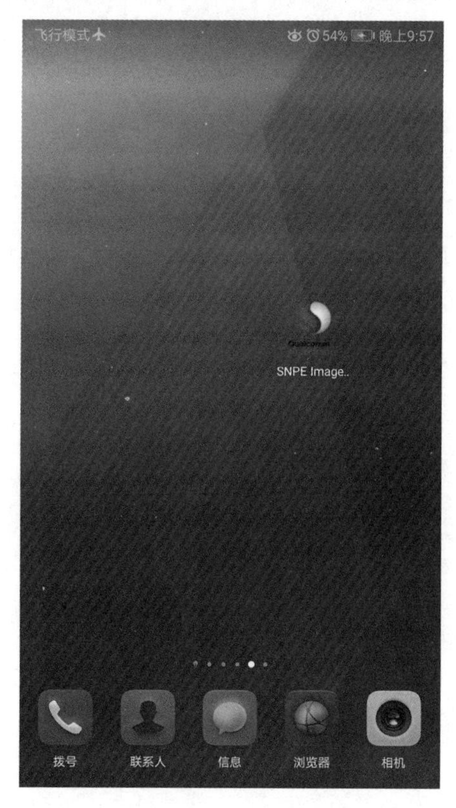

图 9-5　Android 应用图

如果 Android 应用运行在非 Snapdragon 进程中，将会收到如下错误消息：

```
    06-12 08:19:54.001 16812 16812 D SYMPHONY: Detected Symphony running as an Android application, using logcat for all debugging output
    06-12 08:19:54.001 16812 16812 E SYMPHONY: FATAL          0 tef4d04a4 /home/host/build/arm-android-gcc4.9/SecondParty/symphony/src/symphony/src/lib/runtime.cc:468 runtime_init() This version of Symphony is targeted to Snapdragon(TM) platforms
    06-12 08:19:54.001 16812 16812 E SYMPHONY: tef4d04a4 /home/host/build/arm-android-gcc4.9/SecondParty/symphony/src/symphony/src/lib/runtime.cc:468 **********
    06-12 08:19:54.001 16812 16812 E SYMPHONY: tef4d04a4 /home/host/build/arm-android-gcc4.9/SecondParty/symphony/src/symphony/src/lib/runtime.cc:468 - Terminating with exit(1)
    06-12 08:19:54.001 16812 16812 E SYMPHONY: tef4d04a4 /home/host/build/arm-android-gcc4.9/SecondParty/symphony/src/symphony/src/lib/runtime.cc:468 **********
```

以上测试是笔者在小米手机“Qualcomm Snapdragon 660”上做的，下面是这款手机的型号信息：

```
    $ ro.build.fingerprint: [Xiaomi/jason/jason:7.1.1/NMF26X/V9.6.1.0.NCHCNFD:user/release-keys]
```

实例中使用 AlexNet 的图像分类器，我们可以使用不同的硬件，来测量它们的运算速度，以下是测量的结果：

```
    CPU
    time: 292 ms
    time: 304 ms
    time: 272 ms

    DSP
    time: 191 ms
    time: 147 ms
    time: 143 ms

    GPU
    time: 216 ms
    time: 174 ms
    time: 183 ms
```

从以上测试结果可以看到，DSP 表现最佳，GPU 其次。

9.2.2 华为 HiAI Engine

华为为 AI/ML 开发提供了一套开发人员 SDK。同样，用户必须先到开发者网站 http://developer.huawei.com 注册华为 ID。华为开发者首页如图 9-6 所示。

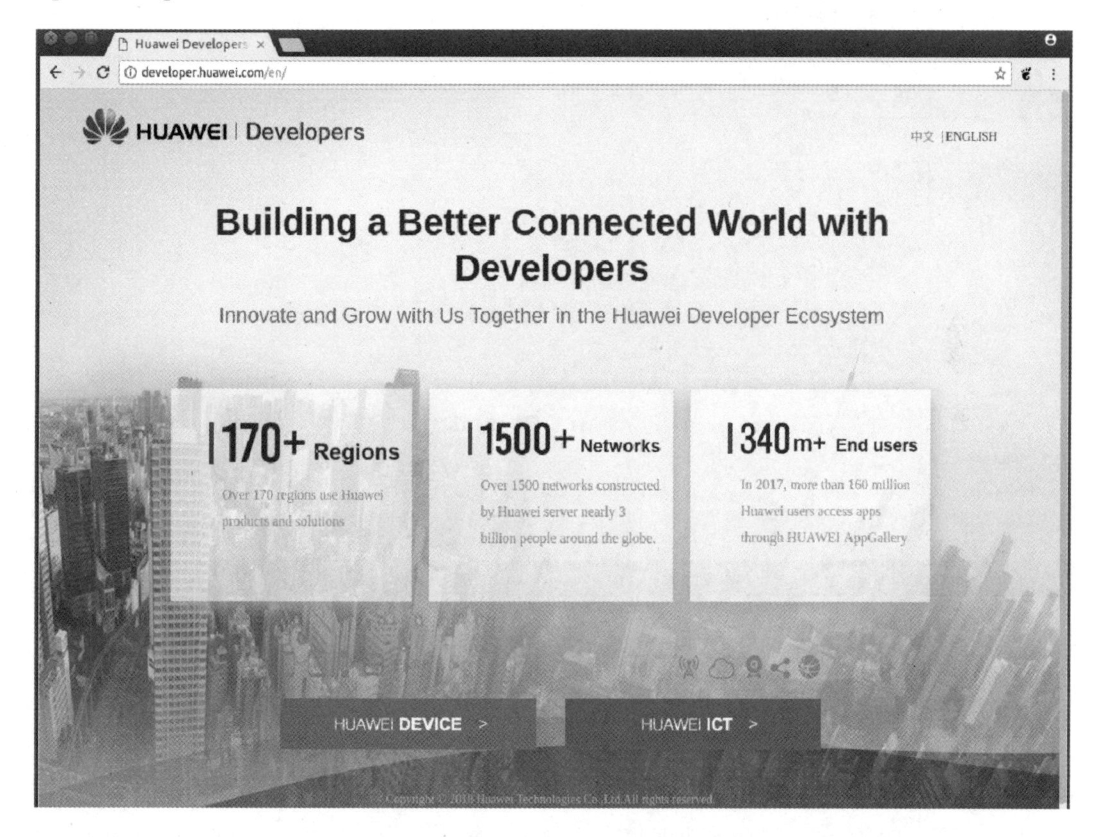

图 9-6 华为开发者首页

登录后，请选择 HUAWEI DEVICE，它是移动设备开发的门户入口。对于个人开发人员来说，注册非常简单，开发者页面如图 9-7 所示。对于公司开发人员，注册的流程是不一样的，在这里就不解释了。

HiAI 是一个面向移动终端的人工智能（AI）计算平台，构建了三层生态：服务能力开放、应用能力开放和芯片能力开放。集成了终端、芯片和云的三层开放平台为用户和开发人员带来了更多的体验。

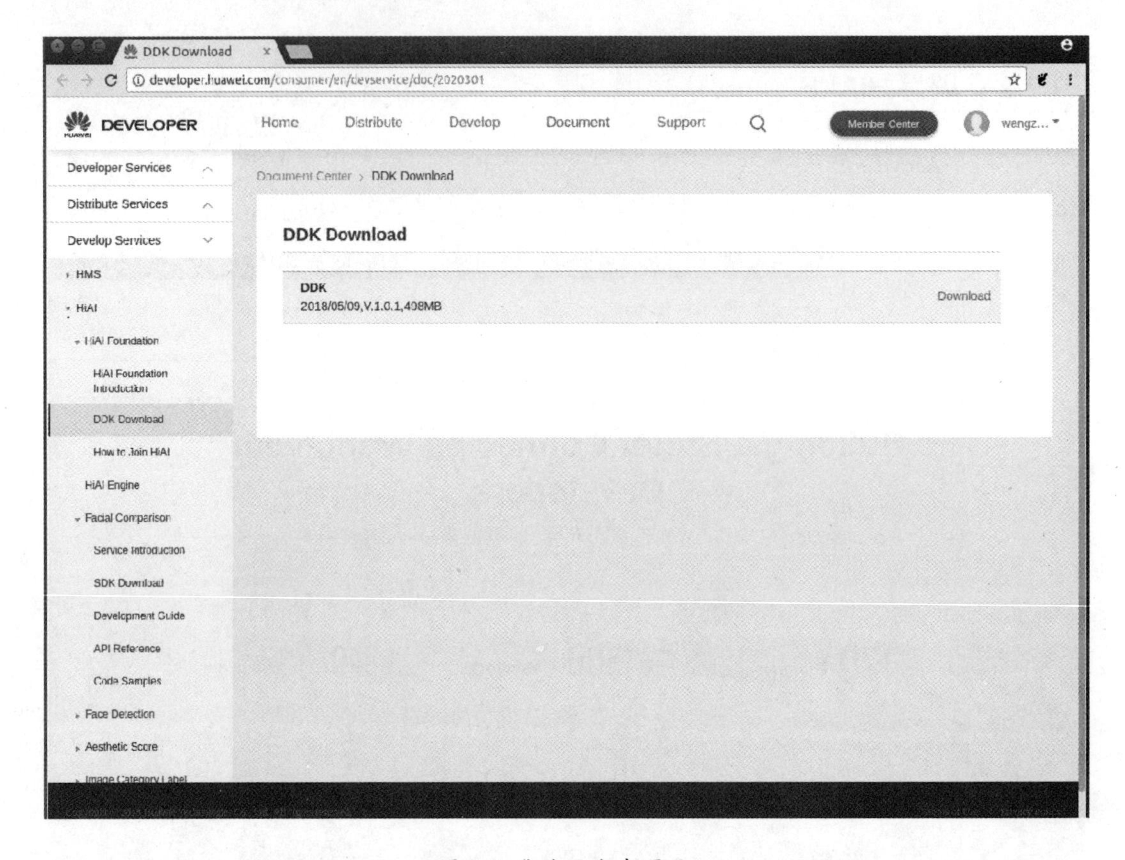

图 9-7 华为开发者页面

华为还提供了 HiAI Engine 编程助手。HiAI Engine 编程助手把所有的 API 以卡片的形势呈现，卡片的详情介绍中有针对 API 的应用场景说明、接口说明和示例代码，直接拖曳卡片到应用的工程中，就会生成 API 调用的 20 多行标准代码，对于更个性化的需求，开发者只需对接口中的参数稍作修改即可，比起手工输入代码的方式，其效率要高很多，而且不容易出错。

另外，HUAWEI DevEco IDE 还提供 7×24 小时远程真机调试服务，能直接连接到华为 Openlab 实验室的真机，实现应用功能远程调测及应用安装与操作。

华为提供了如下 15 个用例演示，这些几乎涵盖了当前市场中你能想到的所有用例。

（1）Facial Comparison

（2）Face Detection

（3）Aesthetic Score

（4）Image Category Label

（5）Image Super-Resolution

（6）Scene Detection

（7）Character Image Super-Resolution

（8）Document Correction/Detection

（9）Code Detection

（10）Face Attributes

（11）Face Orientation Recognition

（12）Face Parsing

（13）Facial Feature Detection

（14）Image Semantic Segmentation

（15）Portrait Segmentation

对于每个用例，开发者网站都提供了 SDK/DDK 和示例代码。SDK 一般提供高级 Java 接口。设备开发工具包（DDK）提供底层 C++接口。华为 HiAI 的开发流程如图 9-8 所示。

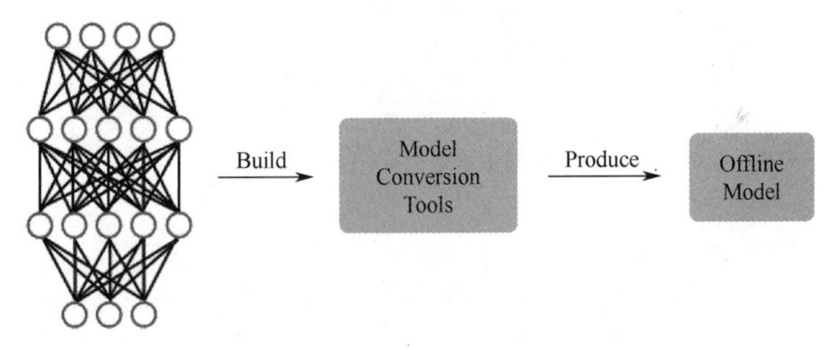

图 9-8　华为 HiAI 的开发流程

如图 9-9 所示说明了华为 HiAI 从 Offline 到硬件加速的开发运行流程。

图 9-9　华为 HiAI 的开发运行流程

HiAI 支持 TensorFlow 和 CAFFE2 模型，我们可以通过名为 cngen 的工具将 TensorFlow 或 CAFFE2 模型转换为 HiAI 引擎支持的格式。

用户文档中提到了“在线模型”和“离线模型”这两个术语，它们是指设备状态。在“离线”模型中，我们不必在设备和云之间建立连接，云服务器首先将模型推送到设备，然后设备自行运行该模型。“在线模型”意味着在推理或培训中，设备和云之间存在连接和数据交换。

下面给出一个运行 DDK 的应用例子，可以让读者比较直观地理解底层的功能。如果读者只关注构建应用，那么了解 SDK 就够了。

我们从华为网站下载 DDK 的代码，可以使用 Gradle 安装，也可以使用 Bazel 构建，编译执行文件的代码如下：

```
    android_library(
        name = "demo_lib",
        srcs = glob([
            "java/**/*.java",
        ]),
        custom_package = "com.huawei.hiaidemo",
        manifest = "AndroidManifest.xml",
        resource_files = glob(["res/**"]),
        deps = [
            "//app/src/main/libs:android_tensorflow_inference_jar",
            "//app/src/main/libs:libhiai",
            "//app/src/main/libs:libtensorflow_inference",
            "//third_party:android_arch_core_common",
            "//third_party:android_arch_lifecycle_common",
            "//third_party:com_android_support_constraint_constraint_
layout_solver",
            "//third_party:com_android_support_support_annotations",
            "@android_arch_lifecycle_runtime//aar",
            "@com_android_support_appcompat_v7//aar",
            "@com_android_support_cardview//aar",
            "@com_android_support_constraint_constraint_layout//aar",
            "@com_android_support_design//aar",
            "@com_android_support_recyclerview//aar",
```

```
        "@com_android_support_support_compat//aar",
        "@com_android_support_support_core_ui//aar",
        "@com_android_support_support_core_utils//aar",
        "@com_android_support_support_fragment//aar",
        "@com_android_support_support_v4//aar",
        "@com_android_support_support_vector_drawable//aar",
    ],
)

android_binary(
    name = "demo",
    custom_package = "com.huawei.hiaidemo",
    assets = [
        "//app/src/main/assets:InceptionV3.cambricon",
        "//app/src/main/assets:inceptionv3_cpu.pb",
        "//app/src/main/assets:labels.txt",
    ],
    assets_dir = "",
    manifest = "AndroidManifest.xml",
    manifest_values = {
        "applicationId": "com.huawei.hiaidemo",
    },
    multidex = "native",
    resource_files = glob(["res/**"]),
    deps = [
        ":demo_lib",
    ],
)
```

华为提供预编译的库文件作为 HiAI 的接口，下面我们重新编译库文件：

```
package(default_visibility = [
    "//visibility:public",
])

cc_import(
  name = "libai_client.so",
```

```
  shared_library = "arm64-v8a/libai_client.so",
)
```

然后，就可以构建执行文件了：

```
$ bazel build app/src/main:demo
```

命令执行后得到应用文件，下面是在手机上安装应用之后的截屏，如图 9-10 所示。

图 9-10　华为演示应用安装图

该应用运行时的截屏如图 9-11 所示。

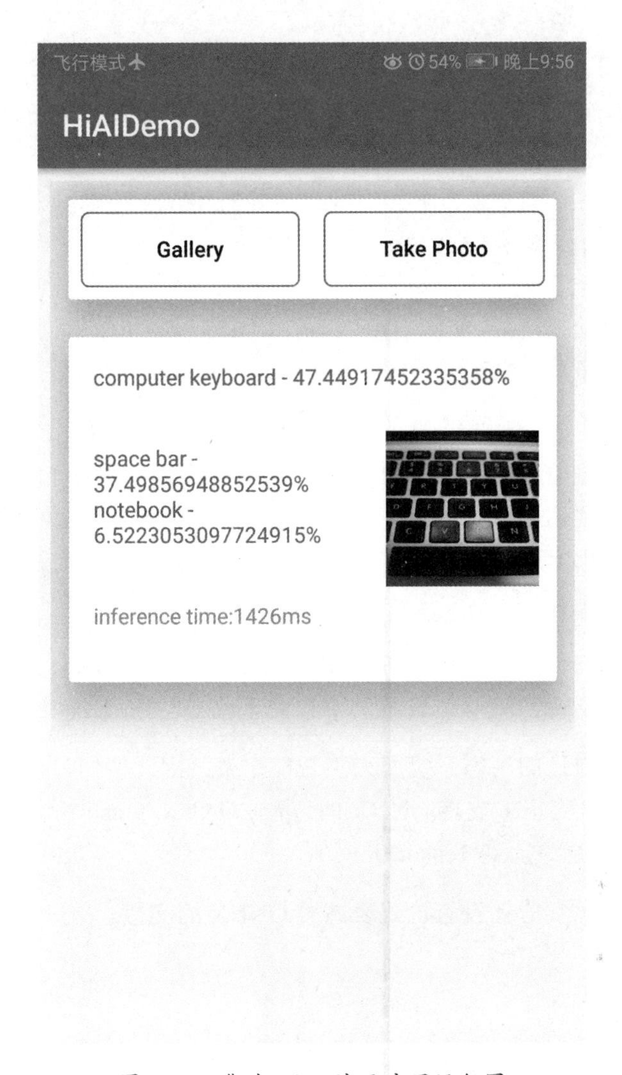

图 9-11　华为 HiAI 演示应用运行图

9.2.3　简要比较

高通 SDK 更专注于开发人员的具体的开发工作。例如，参考指南文档在解释如何有效使用其 API 方面花费了大量的文字。而华为 SDK 更专注于帮助开发人员开发不同类型的应用程序。

例如，华为提供了大量的演示和 Android Studio 插件，希望开发者可以用最小的资源开发出应用。高通可能需要开发者有相应的领域知识，希望开发者能够构建应用。华为SDK则更多是面向普通移动开发者，比较容易上手，笔者个人希望华为的开发包可以做得再精致一些。

9.2.4 开放式神经网络交换格式

开放式神经网络交换格式，简称ONNX，是由Facebook和微软创建的社区项目。ONNX在主页（https://onnx.ai）上声称：

我们认为AI工具社区需要更强的互操作性。许多人正在研究出色的工具，但开发人员经常被锁定在一个框架或生态系统中。ONNX是允许更多这些工具协同工作的第一步，允许它们共享模型。我们的目标是让开发人员能够为他们的项目。使用正确的工具组合。我们希望每个人都能够尽快将人工智能从研究变为现实，而无需工具链的人为摩擦。我们希望您能加入我们的使命！

除了Facebook，微软、亚马逊、联发科、AMD、华为、ARM、IBM、英特尔和高通等一些大公司也宣布支持ONNX。谷歌和苹果尚未就ONNX支持发表官方通知。

谷歌和苹果拥有自己的生态系统和人工智能框架，因此他们可能对支持ONNX不感兴趣。TensorFlow社区有很多支持ONNX的声音。ONNX-TensorFlow是一个GitHub开源项目，它将ONNX支持添加到TensorFlow中。

现在，TensorFlow官方还没有正式表态对ONNX的支持。

10

第 10 章 机器学习应用框架

本章将介绍两个机器学习的应用框架，它们都是谷歌近期推出的比较有影响力的机器学习的服务框架。

10.1 ML Kit

ML Kit 以强大且易于使用的方式为移动开发人员提供了谷歌的机器学习专业知识和服务。机器学习已经成为移动开发中的一种不可或缺的工具，2018 年 Google I/O 上，谷歌推出了测试版 ML Kit。

ML Kit 是一款全新的 SDK，它以 Firebase 上一个功能强大但易于使用的软件包形式，将谷歌的机器学习的能力带给广大的移动开发者。目前，由于谷歌服务不是很容易连通，

Firebase 作为谷歌服务的一部分，在国内直接使用的可能性很低。谷歌也考虑到这种现状，希望以后可以找到一个可以解决的方案。

Firebase 提供了一套移动端的服务，使用起来非常简便。通过简单易用的 API 提供机器学习服务，这是谷歌为降低机器学习门槛所做的努力。ML Kit 的官方网页是 https://developers .google.com/ml-kit。

使用机器学习对许多开发者来说可能非常困难。通常，新的机器学习开发者需要花费大量的时间学习实现底层模型和使用框架等众多复杂操作。即使是经验丰富的专家，修改和优化模型并使其在移动设备上运行也可能是一项艰巨的任务。除机器学习的复杂性外，寻找训练数据也可能是一个费时费力的过程。可是借助 ML Kit，读者可以使用机器学习在 Android 和 iOS 上构建各种应用。

ML Kit 有如下三个特性：

- 对移动设备高度优化。
- 依托谷歌能力。
- 可行性和全面性。

ML Kit 可以使开发者的应用程序更具吸引力、个性化和更有用，并提供针对在移动设备上运行而进行优化的解决方案。ML Kit 提供的技术是谷歌长期以来在移动设备积累的体验。读者根据特定需求，可以使用开箱即用的解决方案（使用基本 API），构建在设备上运行或在云中运行的自定义模型。

ML Kit 通过一个简单的接口为开发者提供设备上的 API 和云 API，读者可以根据自己的需求选择最合适的接口。设备上的 API 可以快速处理数据，甚至可以在没有网络连接的情况下工作；而基于云的 API 则充分利用了 Google Cloud Platform 的机器学习技术，可以提供更高的准确性。

谷歌建议，如果开发者使用 Firebase，需要通过 Firebase 管理部署服务基础架构，那么 ML Kit 会更适合。ML Kit 提供了一组基本 API，并在 TensorFlow Lite 上层为自定义模型包装了薄薄一层。但是，如果开发者没有使用 Firebase 的经验，或没有拥有自定义模型，并希望更多地控制部署/服务等，谷歌强烈建议开发者直接使用 TensorFlow Lite。

ML Kit 的开发是在 TensorFlow 和 TensorFlow Lite 之后，作为一个独立的、为开发者提供机器学习解决方案的软件服务而开发的。对于熟悉机器学习并且有机器学习部署经验的读者，可以直接使用第三方的开发软件进行开发，对于不熟悉机器学习的读者，或者想更快地推出机器学习产品的开发者和公司，可以考虑使用类似于 ML Kit 的开发软件。国内也有类似的为开发者提供机器学习的软件和服务，ML Kit 的长处可以说是依托于谷歌的服务，并且和 TensorFlow 深度整合。

如果读者有丰富的机器学习经验，并发现 ML Kit API 没有涵盖的用例，则可以利用 ML Kit 部署自己的 TensorFlow Lite 模型。读者只需通过 Firebase 控制台上传模型，ML Kit 将负责托管工作，并将它们提供给应用的用户。这样一来，读者可以让模型独立于自己的 APK/软件包，从而减少应用的安装大小。

另外，由于 ML Kit 可以动态提供开发者的模型，所以我们可以更新模型，而不必重新发布应用。

ML Kit 未来还会添加更多功能。但随着应用的功能越来越多，它们也随之增大，这会影响用户从应用商店安装应用的速度，并且可能让用户支付更多的流量费。机器学习会进一步加剧这种趋势，因为模型的大小很容易就能达到数十兆字节。

因此，谷歌决定开发模型压缩功能。具体来说，谷歌正在试验一项功能，它可以让开发者在上传完整的 TensorFlow 模型和训练数据后，获得压缩后的 TensorFlow Lite 模型。希望谷歌可以在近期推出类似的服务。

对于常见的应用用例，ML Kit 为开发者提供了如下示例，并快速投入产品。

- 面部检测。
- 文本识别。
- 条形码扫描。
- 图像标记。
- 地标识别。

我们使用 ML Kit API，只需将数据传入 ML Kit，SDK 就会迅速得到机器学习的结果。这里笔者只是简要地介绍这些例子，更详细的技术文档可以参照 ML Kit 的官方网站。

在使用 ML Kit 之前，我们介绍一下使用的步骤。

第一步，激活 Firbase。有关 Firebase 的具体操作内容，读者可以参照 https://firebase.corp.google.com。如果读者使用 Android Studio，请先安装 Firebase 的插件，使用

SDK Mananger 安装最新的 SDK 后，在 Tool 的菜单栏可以看到如图 10-1 所示的 Firebase 插件选项。

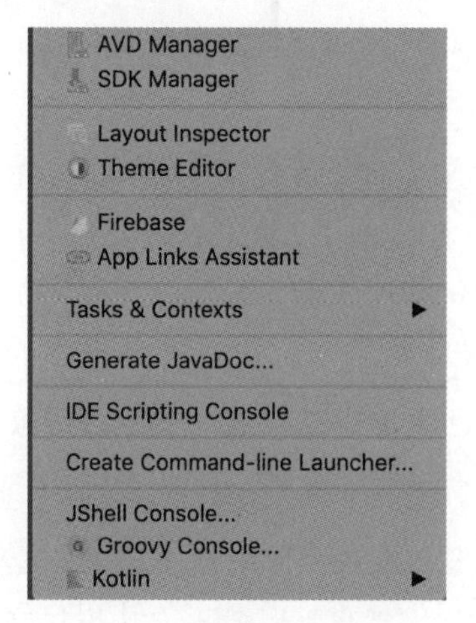

图 10-1　Firebase 插件

单击图 10-1 中的 Firebase 选项，在 Android Studio 里出现 Firebase 的提示页面，如图 10-2 所示。

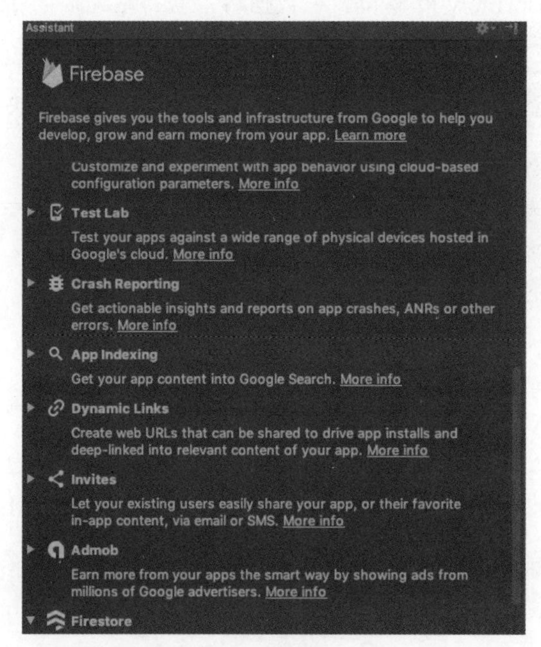

图 10-2　Firebase 提示页面

第二步，注册和激活 Firebase 工程。下面是笔者的 Firebase 主页，如图 10-3 所示。

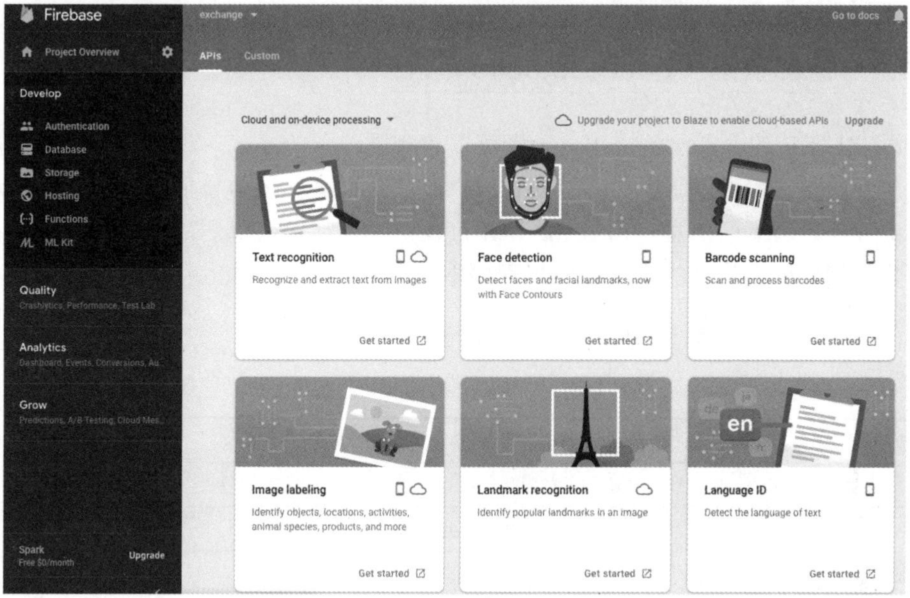

图 10-3　Firebase 主页

单击图 10-3 中右上角的 Upgrade 链接，来到激活页面，如图 10-4 所示。

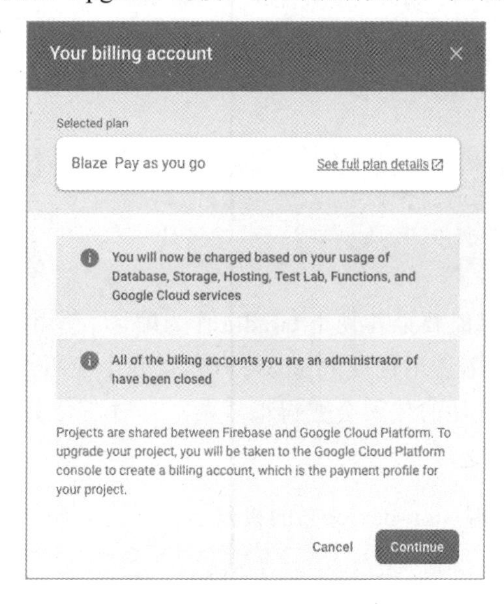

图 10-4　Firebase 激活页面

这两步完成之后，就可以开始使用和开发机器学习的应用了。

10.1.1 面部识别（Face Detection）

使用 ML Kit 的面部检测 API，读者可以检测图像中的面部，识别关键面部特征，并获取检测到的面部轮廓。通过面部检测，用户可以获得执行诸如自拍和肖像或从用户照片生成头像等任务所需的信息。由于 ML Kit 可以实时执行面部检测，因此读者可以在视频聊天或响应播放器表达式的游戏等应用程序中使用它。

面部识别主要特性如表 10-1 所示。

表 10-1 面部识别主要特性列表

特　性	具体内容
识别并定位面部特征	获取检测到的每张脸的眼睛、耳朵、脸颊、鼻子和嘴巴的坐标
获得面部轮廓	获取检测到的面部及其眼睛、眉毛、嘴唇和鼻子的轮廓
识别面部表情	确定一个人是在微笑还是闭着眼睛
跟踪视频帧中的面孔	获取检测到的每个人的面部的标识符，此标识符在调用之间保持一致，因此可以对视频流中的特定人员执行图像处理
实时处理视频帧	面部检测在设备上执行，并且足够快，以满足实时应用，例如视频操作

在实现应用和构建代码之前，我们首先要给应用工程添加依赖。在 build.gradle 里添加下面的依赖：

```
dependencies {
  implementation 'com.google.firebase:firebase-ml-vision:18.0.2'
  implementation 'com.google.firebase:firebase-ml-vision-face-model:17.0.2'
}
```

对于 Firebase 工程，可能要推荐使用 Gradle 作为编译。使用 Bazel 技术上也是可能的，谷歌内部也是使用 build。笔者用也用 Firebase 开发了几个应用，但是，由于 Firebase 本身有第三方的依赖，build 文件的依赖会变得很复杂，有时还会引发一些错误。为了快速开发，用 Gradle 是个可行的选项。

如果读者想实现设备端（on-device）的机器学习，需要添加下面的设定：

```
<application ...>
  <meta-data
```

```
        android:name="com.google.firebase.ml.vision.DEPENDENCIES"
android:value="ocr" />
     <!-- To use multiple models: android:value="ocr,model2,model3" -->
   </application>
```

接下来讲一下图像输入规范。

要使 ML Kit 精确检测面部，输入图像必须包含由足够像素数据表示的面。通常，要在图像中检测的每个脸部应至少为 100×100 像素。如果要检测面部轮廓，ML Kit 需要更高分辨率的输入：每个脸部应至少为 200×200 像素。

如果要在实时应用程序中检测面部，则可能还需要考虑输入图像的整体尺寸，以便能更快地处理较小的图像，从而减少延迟，以较低的分辨率捕获图像并确保主体的脸部尽可能多地占据图像。图像焦点不佳会影响精确度。如果读者没能获得可接受的结果，可以尝试让用户重新捕获图像。如表 10-2 所示是各种功能选项的 API 列表。

表 10-2　各种功能选项的 API 列表

API	函　数	功　能
FirebaseVisionFaceDetectorOptions	build()	构建一个实例
FirebaseVisionFaceDetectorOptions.Builder	enableTracking()	启用面部跟踪，在处理连续帧时为每个面部保持一致的 ID
FirebaseVisionFaceDetectorOptions.Builder	setClassificationMode(int classificationMode)	指示是否运行其他分类器来表征“微笑”和“睁眼”等属性
FirebaseVisionFaceDetectorOptions.Builder	setContourMode(int contourMode)	设置是否检测无轮廓或所有轮廓
FirebaseVisionFaceDetectorOptions.Builder	setLandmarkMode(int landmarkMode)	设置是否检测没有地标或所有地标
FirebaseVisionFaceDetectorOptions.Builder	setMinFaceSize(float minFaceSize)	设置所需的最小面部大小，表示为头部宽度与图像宽度的比例
FirebaseVisionFaceDetectorOptions.Builder	setPerformanceMode(int performanceMode)	扩展选项，用于控制执行面部检测时的额外精度/速度权衡

此外，我们还要设定脸部识别的选项，如表 10-3 所示。

表 10-3　脸部识别的设定选项

设 定 项	具体内容	说　　明
性能模式	FAST (default)	检测面部时有利于速度或准确度
检测地标模式	NO_LANDMARKS (default)	是否尝试识别面部“地标”：眼睛、耳朵、鼻子、脸颊、嘴巴等
检测轮廓	NO_CONTOURS (default)	是否检测面部特征的轮廓。仅针对图像中最突出的面部检测轮廓
脸部分类	NO_CLASSIFICATIONS (default)	是否将面孔分类为“微笑”和“睁眼”等类别
脸部最小大小	float (default: 0.1f)	要检测的面部的最小尺寸，相对于图像
激活脸部追踪	false (default)	是否为面部分配 ID，可用于跟踪图像中的面部

以下是设定对应的代码，

```
    // 高精度地标检测与人脸分类
    FirebaseVisionFaceDetectorOptions highAccuracyOpts =
        new FirebaseVisionFaceDetectorOptions.Builder()
                .setPerformanceMode(FirebaseVisionFaceDetectorOptions.ACCURATE)
                .setLandmarkMode(FirebaseVisionFaceDetectorOptions.ALL_LANDMARKS)
                .setClassificationMode(FirebaseVisionFaceDetectorOptions.ALL_CL
ASSIFICATIONS)
                .build();

    // 多面实时轮廓检测
    FirebaseVisionFaceDetectorOptions realTimeOpts =
        new FirebaseVisionFaceDetectorOptions.Builder()
                .setContourMode(FirebaseVisionFaceDetectorOptions.ALL_CONTOURS)
                .build();
```

下面来看看如何实现脸部识别。要检测图像中的脸部，请从 Bitmap、media.Image、ByteBuffer 或设备上的文件创建 FirebaseVisionImage 对象。然后，将 FirebaseVisionImage 对象传递给 FirebaseVisionFaceDetector 的 detectInImage 函数。对于面部识别，应使用尺寸至少为 480x360 像素的图像。如果要实时识别面部，以最低分辨率捕获帧可以帮助减少延迟。由于相机镜头的转向对识别的结果有影响，所以我们要首先获得相机的转向，下面是对应的代码：

```
    // 获取设备相对于其“本机”方向的当前旋转，然后从“方向”表中查找图像必须达到的角度，旋
转以补偿设备的旋转
        int                          deviceRotation                         =
```

```
activity.getWindowManager().getDefaultDisplay().getRotation();
        int rotationCompensation = ORIENTATIONS.get(deviceRotation);


    // 在大多数设备上，传感器方向为 90 度，但有些设备是 270 度。传感器方向为 270 度，再将图像旋转 180 ((270+270) %360) 度
        CameraManager cameraManager = (CameraManager) context.
getSystemService(CAMERA_SERVICE);
        int sensorOrientation = cameraManager
                .getCameraCharacteristics(cameraId)
                .get(CameraCharacteristics.SENSOR_ORIENTATION);
        rotationCompensation = (rotationCompensation + sensorOrientation + 270)
% 360;
```

然后，生成一个 FirebaseVisionImage 的实例，代码如下：

```
    FirebaseVisionImage image = FirebaseVisionImage.fromMediaImage
(mediaImage, rotation);
```

接着，设定元数据（Meta Data），告诉检测器图像的分辨率、格式和相机的转向，代码如下：

```
    FirebaseVisionImageMetadata metadata = new FirebaseVisionImageMetadata.
Builder()
        .setWidth(480)   // 480×360 通常足以进行图像识别
        .setHeight(360)
        .setFormat(FirebaseVisionImageMetadata.IMAGE_FORMAT_NV21)
        .setRotation(rotation)
        .build();
```

另外，还要生成一个脸部检测器的实例，代码如下：

```
    FirebaseVisionFaceDetector detector = FirebaseVision.getInstance()
        .getVisionFaceDetector(options);
```

最后，调用检测的功能，获得结果 FirebaseVisionFace，代码如下：

```
    Task<List<FirebaseVisionFace>> result =
        detector.detectInImage(image)
                .addOnSuccessListener(
                        new OnSuccessListener<List<FirebaseVisionFace>>() {
                            @Override
                            public void onSuccess(List<FirebaseVisionFace> faces) {
                                // 任务已成功地完成
                            }
```

```
                    })
            .addOnFailureListener(
                    new OnFailureListener() {
                        @Override
                        public void onFailure(@NonNull Exception e) {
                            // 任务失败，出现异常
                        }
                    });
```

至此，脸部识别已经完成。下面笔者通过代码来示范如何正确使用脸部检测结果FirebaseVisionFace：

```
    for (FirebaseVisionFace face : faces) {
        Rect bounds = face.getBoundingBox();
        float rotY = face.getHeadEulerAngleY();  // 头部旋转到rotY度
        float rotZ = face.getHeadEulerAngleZ();  // 头部侧倾，旋转角度

    // 如果启用了Landmark检测（口、耳、眼、脸颊和可用鼻子）：
            FirebaseVisionFaceLandmark leftEar =
                face.getLandmark(FirebaseVisionFaceLandmark.LEFT_EAR);
        if (leftEar != null) {
            FirebaseVisionPoint leftEarPos = leftEar.getPosition();
        }

        // 如果启用轮廓检测：
        List<FirebaseVisionPoint> leftEyeContour =
     face.getContour(FirebaseVisionFaceContour.LEFT_EYE).getPoints();
        List<FirebaseVisionPoint> upperLipBottomContour =
     face.getContour(FirebaseVisionFaceContour.UPPER_LIP_BOTTOM).getPoints();

        // 如果启用了分类：
        if  (face.getSmilingProbability()  !=  FirebaseVisionFace.UNCOMPUTED_
PROBABILITY) {
            float smileProb = face.getSmilingProbability();
        }
        if   (face.getRightEyeOpenProbability()   !=    FirebaseVisionFace.
UNCOMPUTED_PROBABILITY) {
            float rightEyeOpenProb = face.getRightEyeOpenProbability();
        }

        // 如果已启用人脸跟踪：
```

```
    if (face.getTrackingId() != FirebaseVisionFace.INVALID_ID) {
        int id = face.getTrackingId();
    }
}
```

检测器的主要模块 detector.detectInImage 使用了回调函数，人脸识别是在非主线程上完成的。官方网站还提供了 Kotlin 和 iOS 的例子，这里我们就不详述了。Firebase 同时提供了 on-device 和云端机器学习的支持，读者可以根据需求去选择。

10.1.2　文本识别

下面我们来实现文本识别的应用。

首先，在应用里添加依赖：

```
dependencies {
  implementation       'com.google.firebase:firebase-ml-natural-language:
18.1.1'
  implementation       'com.google.firebase:firebase-ml-natural-language-
language-id-model:18.0.2'
}
```

然后，生成 FirebaseLanguageIdentification 的实例，并调用函数去识别文字，实现代码如下：

```
FirebaseLanguageIdentification languageIdentifier =
    FirebaseNaturalLanguage.getInstance().getLanguageIdentification();
languageIdentifier.identifyLanguage(text)
  .addOnSuccessListener(
      new OnSuccessListener<String>() {
        @Override
        public void onSuccess(@Nullable String languageCode) {
          if (languageCode != "und") {
            Log.i(TAG, "Language: " + languageCode);
          } else {
            Log.i(TAG, "Can't identify language.");
          }
        }
      })
  .addOnFailureListener(
      new OnFailureListener() {
        @Override
```

```
        public void onFailure(@NonNull Exception e) {
          // 无法加载模型或其他内部错误
        }
      });
```

10.1.3 条形码识别

下面，我们来实现一个条形码识别的应用。同样，要先添加依赖，再生成实例，接着调用功能去实现条形码，代码如下：

```
FirebaseVisionBarcodeDetectorOptions options =
    new FirebaseVisionBarcodeDetectorOptions.Builder()
    .setBarcodeFormats(
            FirebaseVisionBarcode.FORMAT_QR_CODE,
            FirebaseVisionBarcode.FORMAT_AZTEC)
    .build();
```

上面我们介绍了一些基本应用的实现，这些已经可以覆盖大部分人工智能的应用。此外，我们能感受到应用所需的代码量非常少，更重要的是，它可以帮助我们省去复杂的机器学习操作和管理的步骤，而专注于商业逻辑的开发。应用还可以在设备端学习和云端学习之间进行转换，大大提高了应用的适用场景。

在实际应用中，ML Kit 取得了不错的反响。移动端的推测是基于 TensorFlow Lite 的，所以计算的效果和效率还是不错的。各个应用的模型谷歌内部也在使用，应该能提供比较好的用户体验。

今后，ML Kit 还会提供更多的模型。谷歌团队也在提高 TensorFlow Lite 的能力，以及提供更好的移动端体验，希望 ML Kit 会有一个更广阔的应用前景。

10.2 联合学习（Federated Learning）

谷歌 2017 年发布了联合学习（Federated Learning）的论文，网址是 https://arxiv.org/abs/1602.05629，读者有兴趣可以找原文阅读。同时，谷歌也发表了一篇博客，网址是 https://ai.googleblog.com/2017/04/federated-learning-collaborative.html，算是给自己做个宣传，同时也解释一下联合学习。

在论文发表之前，谷歌已经开始了这方面的研究和产品开发，这个技术被应用到谷歌键盘（Gboard）上。这篇论文既是理论上的发表，也是工程实践的总结。联合学习的基本概念我们可以通过图 10-5 来说明。

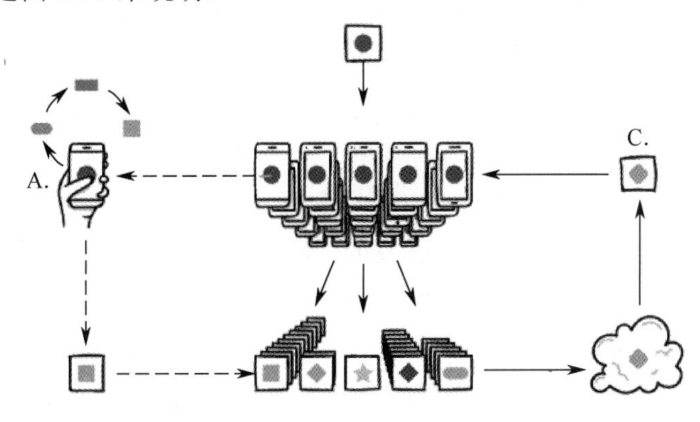

图 10-5　联合学习架构图

联合学习是利用数量庞大的移动设备进行机器学习的一种方法。下面介绍一下联合学习是如何解决数据的隐私问题的。

数据的隐私是一个很重要的问题。从移动端获取用户数据，似乎是最简单最直接的方法。可是在美国，比如在谷歌内部，数据保密和隐私是非常敏感的问题，近几年来，数据隐私变得越来重要。没有用户的同意，任何产品和服务是不能获得用户数据的。

中国的企业很重视用户的隐私。联合学习是用数学的方法解决了这个问题。具体地讲，就是移动端不会向后台或第三方传送用户数据，用户数据会永远留在设备里，机器学习的过程完全是基于移动设备的（on-device）。

机器学习会把训练后的模型传到后端，后端把移动端传送来的模型数据进行汇总和再处理，包括再训练和评价，进而生成一个优化的模型，这个模型会被再次推送到移动端。为了实现这个目的，还有几个问题需要解决。移动端的每个设备都是相对独立的，这和基于云端的分布式训练有很大的不同。这意味着，移动端的机器学习训练是不可预期的，比如训练发生的时间不确定，虽然我们可以通过程序设定训练的频率和开始时间，但是用户随时可以关机或者改变设备状态而引起一些训练时间的改变。

机器学习的训练在移动设备上运行是非常具有挑战性的，有时训练的结果和过程都存在很大的不确定性。联合学习需要设备端把训练后的结果传回后端，但是，由于通信和连接的问题，很难保证设备按照预定的时间表把结果传回来。所以，后端要对接收到的模型参数进行处理。

移动端把训练后的模型参数传到后台后，即使模型的参数泄露，通过对这些参数的分析，我们还是能得到模型的大概信息，通过反向过程，甚至可以推出用户数据。联合学习使用了安全聚合协议（Secure Aggregation Protocol ）来处理这个问题。

联合学习的原理是由一系列论文构成的，这些论文涉及移动端机器学习的各个方面。虽然有了这些理论，实践起来却一点也不轻松。主要原因是，在移动端上进行机器学习和在数据中心或以台式机为主进行机器学习有很大的不同，可能要分成几个小的团队来集体开发这套系统，其中包括一个负责移动开发的团队，一个负责后台的团队。整个系统还要包括监测，以及对这个系统的管理等，工程量还是很大的。

联合学习的流程如图 10-6 所示。

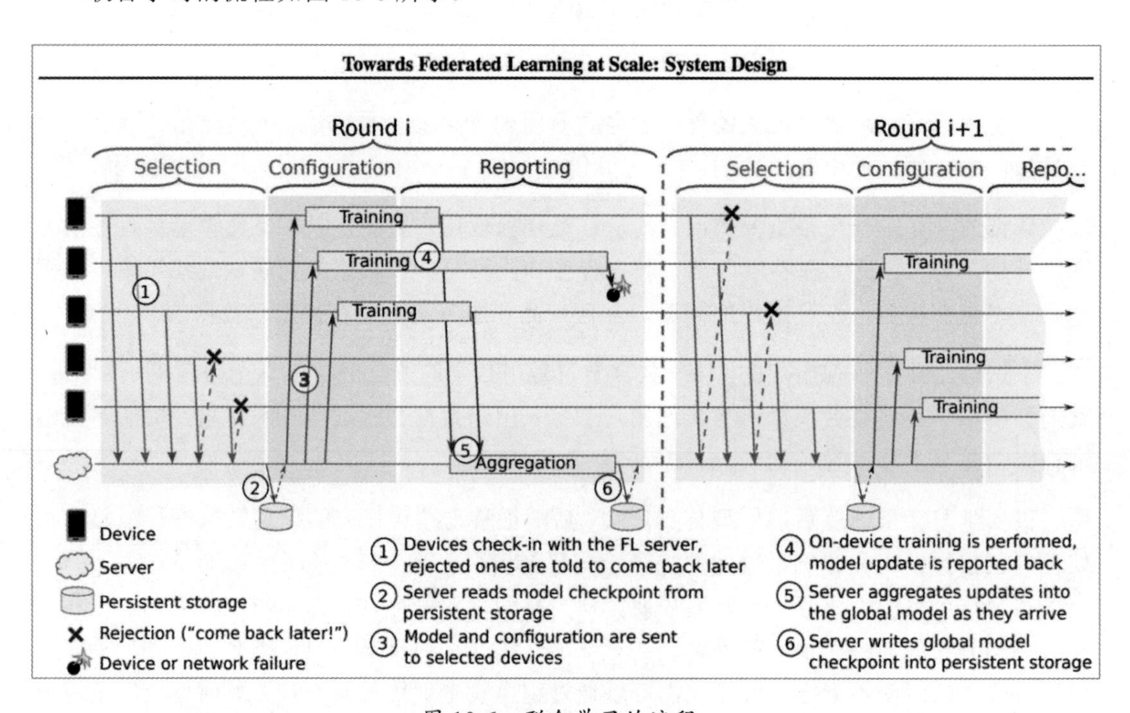

图 10-6 联合学习的流程

联合学习的基本算法如图 10-7 所示。

Algorithm 1 `FederatedAveraging` **targeting updates from K clients per round.**

Server executes:
　initialize w_0
　for each round $t = 1, 2, \ldots$ **do**
　　Select $1.3K$ eligible clients to compute updates
　　Wait for updates from K clients (indexed $1, \ldots, K$)
　　$(\Delta^k, n^k) = \text{ClientUpdate}(w)$ from client $k \in [K]$.
　　$\bar{w}_t = \sum_k \Delta^k$　*// Sum of weighted updates*
　　$\bar{n}_t = \sum_k n^k$　*// Sum of weights*
　　$\Delta_t = \Delta_t^k / \bar{n}_t$　*// Average update*
　　$w_{t+1} \leftarrow w_t + \Delta_t$

ClientUpdate(w):
　$\mathcal{B} \leftarrow$ (local data divided into minibatches)
　$n \leftarrow |\mathcal{B}|$　*// Update weight*
　$w_{\text{init}} \leftarrow w$
　for batch $b \in \mathcal{B}$ **do**
　　$w \leftarrow w - \eta \nabla \ell(w; b)$
　$\Delta \leftarrow n \cdot (w - w_{\text{init}})$　*// Weighted update*
　// Note Δ is more amenable to compression than w
　return (Δ, n) to server

图 10-7　联合学习算法

第 11 章 基于移动设备的机器学习的未来

本章我们将了解机器学习最新的一些进展和趋势。

首先来看一下 TensorFlow 的最新动态。

11.1 TensorFlow 2.0 和路线图

TensorFlow 本身是一个快速发展的开源社区支持的项目。接下来简单介绍一下 TensorFlow 2.0 和未来的发展趋势。

TensorFlow 2.0 将会是一个重要的里程碑，这个版本的重点是易用性。以前 TensorFlow 可能只有有经验的开发者才能使用，TensorFlow 2.0 希望能通过降低使用门槛，让更多的开发者使用 TensorFlow，未来能看到更多的人工智能使用场景。

TensorFlow 2.0 的主要特点如下：

- 使用 Keras 快速开发模型。
- 立即执行（Eager Execution）是 2.0 版本的核心功能。这使 TensorFlow 更易于学习和应用。
- 使用跨平台的、更可靠的产品模型发布。
- 通过交换格式的标准化和 API 的一致性，支持更多平台和语言，并改善组件间的兼容性。
- 为研究人员提供更有效的实验平台。
- 具有更简单的 API，删除过时和重复的 API。

在过去几年中，TensorFlow 添加了许多组件。通过 TensorFlow 2.0，这些组件将被打包成一个综合平台，支持从训练到部署的完整的机器学习工作流程。让我们看一下 TensorFlow 2.0 的新架构，在此使用简化的概念图，如图 11-1 所示。

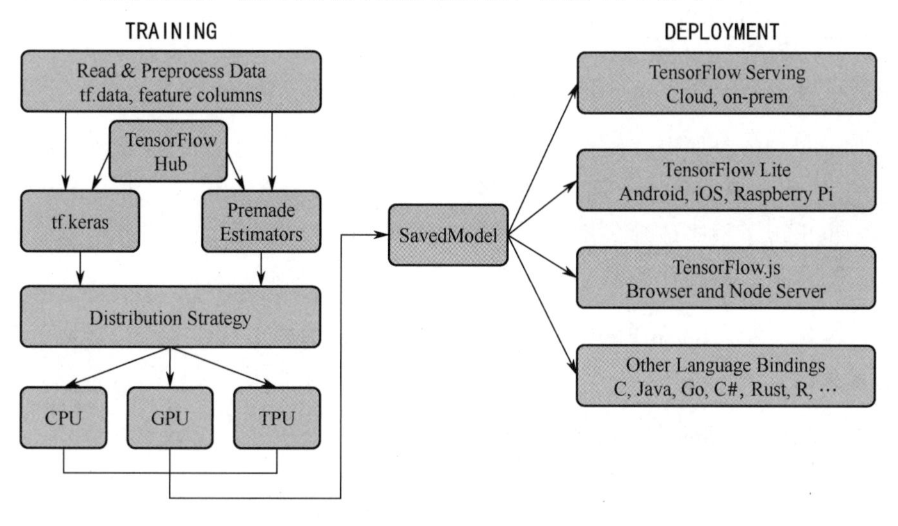

图 11-1　TensorFlow 2.0 架构图

11.1.1　更简单的开发模型

在 TensorFlow 2.0 中，Keras 会成为一个用户友好的机器学习高级 API 的标准，并将成为用于构建和训练模型最重要的高级核心 API。Keras API 让开发者可以轻松开始使用 TensorFlow。重要的是，Keras 提供了一些模型创建 API，让开发者可以为开发工程选择合适的抽象级别。TensorFlow 2.0 的功能还包括 Eager Execution、快速迭代和直接调试，以及用于构建可扩展输入管道的 tf.data。

11.1.2 更可靠的跨平台的模型发布

TensorFlow 2.0 将极大提高跨平台的兼容性，使模型从开发到产品发布更加流畅。2.0 版本的跨平台相关的要点如下：

- 使用 TensorFlow Serving。现在支持 HTTP/REST 和 gRPC。
- 使用 TenforFlow Lite。对于移动设备、IoT 和边缘设备，使用新的 TensorFlow Lite。
- 使用 TensorFlow js。网页端使用 TensorFlow js。

11.1.3 TensorFlow Lite

这里我们重点看一下 TensorFlow Lite。

TensorFlow Lite 是面向移动设备、嵌入设备的默认架构。我们已经看到谷歌内部团体也在大量采用 TensorFlow Lite。同时，很多开发者也在开发基于 TensorFlow Lite 的产品，而且市场上也出现了很多成功的产品。TensorFlow Lite 和 Edge TPU 对软件和硬件搭档会有很好的应用前景，即使作为独立的部件，它们也会各自独当一面。

但是，客观地看，TensorFlow Lite 距离 TensorFlow 的成熟度还有一段距离，而且也面临很大的挑战。

今后 TensorFlow Lite 将要着重发展的方向如下：

- 增加 TensorFlow Lite 中支持的算子的覆盖范围。
- 更容易转换训练过的 TensorFlow 图，以便在 TensorFlow Lite 上使用。
- 加强用于移动模型优化的工具。
- 扩展对 Edge TPU、TPU AIY 开发板的支持。
- 更好的文档和教程。

我们期待更好、更易用和更高效的机器学习框架。

11.1.4　TensorFlow 1.0 和 TensorFlow 2.0 的不同

除了以上 TensorFlow 2.0 的新特性，2.0 版本还清除了一些不合适和过时的 API，并使 API 更加统一。其中一个大的变化是，tf.contrib 将被移除！

Contrib 在 1.0 版本是一个实验产品的集合，由于 TensorFlow 快速普及和被大量采用，Contrib 变得越来越大。从 2.0 版本开始，只有核心功能会加到主干，其他特性和分支会独立出来，尽可能将 tf.contrib 中的大型项目移到单独的代码库中。TensorFlowLite 已经在 2018 年底从 Contrib 里移了出来。

11.2　人工智能的发展方向

本书的最后，笔者想讨论一下人工智能的未来发展方向。既然 TensorFlow 是由谷歌发起的，笔者就整理一下谷歌对人工智能未来发展的一些公开的文章和论点，也会有一些个人看法。

最近几年，随着机器学习和深度学习的发展，人工智能已经深入我们生活的各个方面，有些已经成为生活中不可缺少的一部分。人工智能以后的发展对我们的生活将会有更深远的影响，接下来从三个方面进行探讨：

- 提高人工智能的可解释性。
- 贡献社会。
- 改善生活。

11.2.1　提高人工智能的可解释性

如何能够让人工智能更加透明，让我们更好地理解机器学习的结果，是以后人工智能发展的一个重要方向。就像今天我们仍然无法完全理解人脑的活动和人类的思考方式，机器学习的基本原理也是在模仿人类的思考方式。

许多深度学习算法自诞生起就一直被人们视作“神秘黑匣”，这是因为就连它们的发明者也难以准确表达输入和输出之间究竟发生了什么。如果我们继续把人工智能当作“神秘黑匣”，那么我们就不能期望得到用户和社会的信任，因为信任源自了解。

对于传统软件，我们可以通过逐行检查源代码来揭示其中的逻辑，但神经网络是一个通过数千个乃至数百万个训练示例而形成的密集连接网络，所以很难用已有的方式去理解里面的逻辑关系。这也造成了人们对机器学习的不信任和更多的神秘感。

现在的机器学习，很大程度上还是非常依赖数据的。通过大量数据的训练让机器去理解事物的本质，这是一种有效的方式，也是现在人工智能的主流方式，这和人类的认知方式也很相像。但是，当我们反问自己，我们是否真正理解了这个过程的实际意义？我们真正理解了结果的含义吗？

可能我们并不能完美地回答自己。

科学的发现和工程的进展是一个螺旋上升的过程，被证明是有效的事物，会更加促进我们去了解这个事物的本质。如果我们对人类学习有了更加深入的理解，也许会使机器学习上一个新的台阶。

随着最佳实践的建立，有效而成熟的工具越来越多，再加上大家都在努力从开发周期开始就尽可能取得可解释的结果，这方面的研究正在不断取得进展。

业界已经在实践中思考可解释性的问题。例如，在图像分类领域，谷歌最近的研究演示了一种“人性化表示”的概念，比如图像分类器能够根据对人类用户最有意义的特点来清楚表达其推理过程。一个相关的例子是，图像分类器可能将图像归类为“斑马”，部分原因在于图像中的“条纹”特征较明显，而“圆点花纹”特征相对不够明显。

实际上，研究人员正在试验将这种方法应用于糖尿病视网膜病变的诊断，它可以使输出结果更加透明。当专家不同意模型的推断时，甚至允许对模型进行调整。

11.2.2 贡献社会

如何能够确定机器学习的模型对社会中的每一个个体都有贡献，也是以后研究的一个方向。机器学习越来越成为社会的基础服务和技术架构，我们每天都能接触到机器学习的一些应用。但当我们享受到人工智能带来的服务的时候，我们也要保证身处社会中的每一个人都能受惠于新技术。

当正常人能够使用人工智能服务的时候，我们也要保证盲人和聋哑人等也能够同样使用新技术，分享新技术带来的福利。在谷歌，每个新的产品发布都有一个严格和标准的审查流程，审查中的一项就是这个产品是否可以被有障碍的人群使用，对有障碍人群会不会产生不好的效果。

我们设计的产品，都有自己的特定人群和场景，这无可厚非，但是如果我们能多考虑一些特殊的场景，为儿童、障碍人士、女士设计出更多符合他们生活习惯的产品，也是技术对这个社会做出的贡献。

另一个视角是，即使我们有意识地为全体人群和用户考虑设计产品，但是由于设计的失误和欠缺，也会导致最终的产品产生偏差。比如，我们在收集数据的时候，有没有考虑数据的全面性。如果我们从和我们有共同生活样式的人群里收集数据并设计模型，这个模型很难适用于全体人群。这是非常明显的例子，似乎没人会犯这个错误。但是，如果我们重新设计模型，从有限的数据开始训练模型的时候，一定要考虑模型的偏差，避免相似的错误。

此外，由于我们自己存在偏见，因此即便是如实收集数据，也可能把这种偏见表现出来。例如，大量的历史文本经常用于训练涉及自然语言处理或翻译的机器学习模型，如不修正，可能会使某种有害的成见持续下去。

有研究尝试用清晰度量化的方式来反映这一现象。实际研究证明，统计语言模型能够非常轻松地“学习”关于性别的过时假设。例如“医生”是“男性”，“护士”是“女性”。与此相似的偏见问题在人群族裔方面也有体现。业界正在多个领域解决这些问题，其中以感知领域最为重要。为了促进人们更广泛地理解公平对于机器学习等技术的必要性，业界在教育界投入了大量资源，希望在教育速成课程中推广和普及对公平性的理解。

当然，业界一直致力于为开发者提供值得信赖的工具，在去除偏见方面也是一样。首先从文档开始，例如完善机器学习指南文档，谷歌将此种指南集成在 AutoML 中，并扩展到类似 TensorFlow Model Analysis（TFMA）和 What-If 等的工具里。该指南为开发者提供所需的分析数据，使开发者确信他们的模型会公平对待所有用户。TFMA 可以轻松地将模型在用户群体的不同环境、特征和子集下的性能表现可视化，而 What-If 支持开发者运行反设事实，阐明关键特征（例如给定用户的人口属性）逆转时可能会发生的情况。这两个工具都可以提供沉浸式的互动方法，用于详细探索机器学习行为，帮助我们识别公平性和代表性方面的失误。

对于目前采取的这些做法，笔者相信这些知识和正在开发的工具具有深远的意义，必将促进人工智能技术的公平性。但没有一家公司能够独自解决如此复杂的问题。这场对抗偏见的战斗将是一次全社会的集体行动，由许多利益相关的人和群体投入，共同推动。

11.2.3 改善生活

我们也希望，科学家和工程师能够负责任地利用人工智能和自动化技术，来确保我们能更好地改善生活，并为未来做好准备。自从人工智能被广泛应用之后，我们常常听到的是以后哪些人群、哪些行业会被人工智能所代替，这并不是危言耸听，而是正在发生的事实。

不过，笔者认为，人工智能的未来并非一场零和游戏，或是对人类的毁灭。最近的一份报告也显示，很多企业的高管认为，通过人工智能和人类智能相结合，人工智能将助推人类和机器协同工作，发挥更强大的作用。

另外，我们要知道，工作很少会是单一、简单的。大多数工作都是由无数不同任务组成的，从高度创新到重复性任务，每一项任务都会在特定程度上受到自动化技术的影响。我们希望这种改变能推动社会向好的方向发展，而不是向毁灭的方向发展。

人类的每次技术革命都会产生阵痛，在这次的人工智能变革中，我们也希望技术能减少阵痛对社会的冲击。比如，通过人工智能技术，对儿童进行更高效的教育，对未来从事可被替换工种的人群提供个性化的培训等。

然而，某些类别的工作面对的变化要比其他工作更加剧烈，并且要做出更多努力才能缓和这种转变。但这并不是人类第一次面临这样的挑战和困难。人类历史上第一次和第二次工业革命给社会带来的冲击都非常巨大。

从信息革命以后，人类的技术进步和革新的步伐并没有停止，而是越来越快。有些行业消失了，又有些新的行业诞生了。我们希望通过技术改善生活，同时，正视技术淘汰，正确面对越来越快的改变。在这种改变中，技术应该是推动社会、改善生活的正面角色，而不是人类的敌人。

電子工業出版社 PUBLISHING HOUSE OF ELECTRONICS INDUSTRY http://www.phei.com.cn Broadview® WWW.BROADVIEW.COM.CN

博文视点 · IT出版旗舰品牌

博文视点精品图书展台

专业典藏

移动开发

大数据 · 云计算 · 物联网

数据库

Web开发

程序设计

软件工程

办公精品

网络营销

反侵权盗版声明

电子工业出版社依法对本作品享有专有出版权。任何未经权利人书面许可，复制、销售或通过信息网络传播本作品的行为；歪曲、篡改、剽窃本作品的行为，均违反《中华人民共和国著作权法》，其行为人应承担相应的民事责任和行政责任，构成犯罪的，将被依法追究刑事责任。

为了维护市场秩序，保护权利人的合法权益，我社将依法查处和打击侵权盗版的单位和个人。欢迎社会各界人士积极举报侵权盗版行为，本社将奖励举报有功人员，并保证举报人的信息不被泄露。

举报电话：（010）88254396；（010）88258888

传　　真：（010）88254397

E-mail：　dbqq@phei.com.cn

通信地址：北京市海淀区万寿路 173 信箱

电子工业出版社总编办公室

邮　　编：100036